Kevin Stella

Electronic dissipation processes during chemical reactions on surfaces

disserta
Verlag

Stella, Kevin: Electronic dissipation processes during chemical reactions on surfaces,
Hamburg, disserta Verlag, 2011

ISBN: 978-3-942109-88-8
Druck: disserta Verlag, ein Imprint der Diplomica® Verlag GmbH, Hamburg, 2011

Bibliografische Information der Deutschen Nationalbibliothek
Die Deutsche Nationalbibliothek verzeichnet diese Publikation in der Deutschen
Nationalbibliografie; detaillierte bibliografische Daten sind im Internet über
http://dnb.d-nb.de abrufbar.

Die digitale Ausgabe (eBook-Ausgabe) dieses Titels trägt die ISBN 978-3-942109-89-5
und kann über den Handel oder den Verlag bezogen werden.

Fakultät für Chemie
Physikalische Chemie
Universität Duisburg-Essen
Universitätsstr. 5
D-45141 Essen

Dieses Werk ist urheberrechtlich geschützt. Die dadurch begründeten Rechte, insbesondere die der Übersetzung, des Nachdrucks, des Vortrags, der Entnahme von Abbildungen und Tabellen, der Funksendung, der Mikroverfilmung oder der Vervielfältigung auf anderen Wegen und der Speicherung in Datenverarbeitungsanlagen, bleiben, auch bei nur auszugsweiser Verwertung, vorbehalten. Eine Vervielfältigung dieses Werkes oder von Teilen dieses Werkes ist auch im Einzelfall nur in den Grenzen der gesetzlichen Bestimmungen des Urheberrechtsgesetzes der Bundesrepublik Deutschland in der jeweils geltenden Fassung zulässig. Sie ist grundsätzlich vergütungspflichtig. Zuwiderhandlungen unterliegen den Strafbestimmungen des Urheberrechtes.

Die Wiedergabe von Gebrauchsnamen, Handelsnamen, Warenbezeichnungen usw. in diesem Werk berechtigt auch ohne besondere Kennzeichnung nicht zu der Annahme, dass solche Namen im Sinne der Warenzeichen- und Markenschutz-Gesetzgebung als frei zu betrachten wären und daher von jedermann benutzt werden dürften.

Die Informationen in diesem Werk wurden mit Sorgfalt erarbeitet. Dennoch können Fehler nicht vollständig ausgeschlossen werden und der Verlag, die Autoren oder Übersetzer übernehmen keine juristische Verantwortung oder irgendeine Haftung für evtl. verbliebene fehlerhafte Angaben und deren Folgen.

© disserta Verlag, ein Imprint der Diplomica Verlag GmbH
http://www.disserta-verlag.de, Hamburg 2011
Hergestellt in Deutschland

Electronic dissipation processes during chemical reactions on surfaces

Dissertation
zur Erlangung des Doktorgrades
der Naturwissenschaften
- Dr. rer. nat. -

vorgelegt von

Kevin Stella

geboren am 15. April 1984 in Duisburg

Fakultät für Chemie
Universität Duisburg-Essen

Juni 2011

Die vorliegende Arbeit wurde im Zeitraum von Juni 2008 bis Juni 2011 im Arbeitskreis von Prof. Dr. Eckart Hasselbrink am Institut für Physikalische Chemie der Universität Duisburg-Essen durchgeführt.

Tag der Disputation: 20.09.2011

Gutachter: Prof. Dr. Eckart Hasselbrink
                       Prof. Dr. Marika Schleberger
                       Prof. Dr. Achim W. Hassel

Vorsitzender: Prof. Dr. Mathias Ulbricht

"Your own resolution to succeed is more important than any other thing"
Abraham Lincoln

Contents

1 Introduction   1

2 Basic information   3
    2.1 Adiabatic or non-adiabatic reaction? 3
    2.2 Molecular beams 9
    2.3 Cyclic Voltammetry 14
    2.4 Water formation reaction 18

3 Experimental design and sample preparation   25
    3.1 The Ultra High Vacuum recipient 25
    3.2 The molecular beam apparatus 27
    3.3 The ion beam source 27
    3.4 Tunnel detectors 29
       3.4.1 Metal–insulator–metal tunnel detector (MIM) 29
       3.4.2 The new stepped metal–insulator–semiconductor tunnel detector (stepped–MIS) 30

4 Measurements, results and discussion   33
    4.1 Molecular beam 33
       4.1.1 Translational energy of the particles 33
       4.1.2 Particle numbers of the molecular beam 36
    4.2 Ion beam characteristics 42
    4.3 Ion migration in anodic oxide films after potentiostatic formation . 43
       4.3.1 Introduction 43
       4.3.2 Experimental 46
       4.3.3 Results 49
       4.3.4 Discussion 56
       4.3.5 Conclusion 64
       4.3.6 Appendix 65
    4.4 Preparation and properties of thin amorphous tantalum films . . . 66
       4.4.1 Introduction 66

	4.4.2	Experimental	67
	4.4.3	Electrical resistivity of thin tantalum films	76
	4.4.4	Conclusion	81
4.5	Thin tantalum films on crystalline silicon – a metallic glass		86
	4.5.1	Introduction	86
	4.5.2	Experiment and Discussion	86
	4.5.3	Conclusion	91
4.6	Charge transport through thin amorphous TiOx and TaOx layers	92
	4.6.1	Introduction	92
	4.6.2	Sample preparation	93
	4.6.3	Internal photoemission	97
	4.6.4	Bias-induced charge transport	108
	4.6.5	Chemicurrents	113
	4.6.6	Conclusion	114
	4.6.7	Appendix	116
4.7	Photo-sensitive MIS sensors with stepped insulating layer	118
4.8	Transport of excited holes through MIS sensors	125
	4.8.1	Introduction	125
	4.8.2	Experimental	127
	4.8.3	Electronic sensor properties	129
	4.8.4	Capacitance voltage plots at different thicknesses	135
	4.8.5	Theoretical description of internal photoemission	137
	4.8.6	Experimental results	141
	4.8.7	Conclusion	157
	4.8.8	Mathematical Appendix	158
4.9	Molecular reaction chemicurrent studies with MIS sensors	160
	4.9.1	Study of effusing molecular gases with MIS sensors	160
	4.9.2	Direct dosage of excited species on a MIS sensor I	162
	4.9.3	Direct dosage of excited species on a MIS sensor II	165
	4.9.4	The origin of the chemicurrent signals	166
	4.9.5	Time dependent trace of the chemicurrent signals	175
	4.9.6	Indirect dosage of excited species on a MIS sensor	179
	4.9.7	Effusion of deuterium and oxygen mixtures	184

5 Summary and Outlook   **187**

Bibliography   **193**

6 Further experimental results planned for publication — 217
- 6.1 Temperature dependence of internal photoemission in MIM sensors — 217
- 6.2 Temperature dependence of internal photoemission in MIS sensors — 222
- 6.3 About the stability of thin Al–AlOx and Ta–TaOx interfaces — 225
  - 6.3.1 Introduction — 225
  - 6.3.2 Experimental — 225
  - 6.3.3 Discussion — 226
  - 6.3.4 Conclusion — 231

7 Appendix — 233
- 7.1 Refereed Papers published — 233
- 7.2 Refereed Papers in preparation — 234
- 7.3 Invited talk — 235
- 7.4 Poster and talks — 235

Curriculum vitae — 238

Erklärung — 239

Danksagung — 240

1 Introduction

Every day in our life is larded with a huge number of chemical reactions on surfaces. Some reactions occur immediately, for others an activation energy has to be supplied. Thus it happens that though a reaction should thermodynamically run off, it is kinetically hindered. Meaning the partners react only to the thermodynamically more stable product state within a mentionable time if the activation energy of the reaction is supplied. With the help of catalysts the activation energy of a reaction can be lowered. Such catalytic processes on surfaces are widely used in industry. Around 90 % of chemicals are produced via a heterogeneously catalyzed process where a reaction occurs on the surface of a catalyst [1, 2].

However, why it is generally possible, that such reactions run off with the help of a heterogeneous catalysis, meaning with a lower activation energy than without presence of a catalyst? What happens with the energy, which is released during a reaction of gas particles on surfaces? How is this energy released, when some part of the energy is transferred to the reactant and some to the chemically active surface? Which physical mechanisms play a key role in the energy transfer?

These questions are summarised in the concept of the energy dissipation. To observe these energy dissipation phenomenon, we use a new method, the chemoelectronics. With this method we try to detect the released energy, induced by reactions on surfaces, via thin-layered electronic device elements [3–13].

It is an aim of this work, to build up very sensitive chemoelectronic sensors to measure electronic excitations released during such simple reactions of molecules as adsorption and desorption and more complicated reactions as the water formation reaction. Therefore a new line of chemoelectronic sensors is developed and characterized in terms of internal photoemission and stability. Meaning the previously used aluminum (Al–AlOx–Ag) [14, 15] and tantalum based (Ta–TaOx–Au) [7–9] **metal–insulator–metal** sensors (MIM) are tested and new titanium based (Ti–TiOx–Au) MIMs are developed. Additionally silicon based **stepped–metal–insulator–semiconductor** sensors (stepped-MIS, Si–SiOx–Au, Si–SiOx–Pt) are set up and characterized.

1 Introduction

For the characterization of the chemoelectronic sensors the process of internal photoemission is used. Both, chemical- and photoexciation release hot charge carriers (electrons and holes). Due to the existence of excited carriers in the sensor, a current can be measured without applying a bias voltage. It will be shown that the chemo- and the photosensitivity are strongly related to each other. As a first test experiment for the chemical selectivity of the detectors a stream of excited hydrogen molecules and hydrogen atoms is used. The excitation and radical formation is produced by the interaction of ground state molecules with a hot tungsten surface according to the pioneering experiments of Irving Langmuir [16–19]. Additionally excited oxygen beams are studied in this work.

2 Basic information

2.1 Adiabatic or non-adiabatic reaction?

When an exothermic reaction takes place on a metal surface, the released energy can be transferred into the degrees of freedom of the reaction products or into the lattice of the metal. If this energy is transferred only in lattice vibrations respectively phonons, this process is denoted as an adiabatic process. In this adiabatic process the electrons in the solid and the reacting species remain in the ground state during the energy transfer, since the motion of the reactands nuclei is much slower than the velocity of the electrons.

However, it is also known that in some cases, for example, in the oxidation of magnesium, electrons are excited, because an emission of light can be observed (chemiluminescence) [20–22]. As well, one can observe a so called exoemission, as an emission of electrons into the vacuum occurs. Such an effect of exoelectron emission occurs especially when atoms or molecules of large electronegativity (e.g. oxygen or halogens) interact with low work function metals [10,23,24]. Additionally one can observe the ejection of negative ions resulting from impinging molecules as well as from dissociation fragments when molecules or atoms undergo thermal collisions with surfaces [10,24–27]. These processes belong to the class of non-adiabatic processes. Non-adadiabatic processes may also comprise processes described as electronic friction. In these processes the electron system of the metal gets excited under formation of electron-hole pairs. The formed hot electrons have an energy which is higher than the Fermi energy. The excited or hot hole state (missing electron in the valence band) has an energy that is smaller than the Fermi energy [10].

Let's have a closer look on these effects. As mentioned exothermic chemical reactions carried out on metal surfaces can dissipate a part of their energy into excited electron–hole pairs in the metal surface. This is possible since metals have a continuous electronic spectrum what allows in principle electronic excitations of any quantum [10]. Therefore one can think about the elementary (e.g. adsorption, desorption, dissociation and vibrational damping) steps of a chemical reaction, where each of the elementary steps might excite the surface electronic system, provided that the reactant's motion is fast enough (see below).

However, for semiconductors this is different, since they have a band gap and thus a non-continuous energy spectrum. Therefore just energies larger than the band gap (e.g. for silicon 1.1 eV [28]) can be exchanged between the particles and the surface. Thus such processes are improbable and quite inefficient, leading to nearly no excitation of the semiconductor surface's electronic system.

Due to the linked rearrangement of charges during bond formation and breaking, an excited electronic state of the substrate can occur if the these processes are fast. Meaning a nearly instantaneous respond of the metal substrate electrons occurs [10]. To sum up in words of this review by Hasselbrink, three origins of non-adiabaticity exist: 1) the mentioned delayed charge transfer, 2) an oscillating charge transfer between states derived from molecular orbitals and the density of states of the substrate close to the Fermi level and 3) slow spin polarization. Furthermore Hasselbrink denotes that non-adiabatic effects are prevalent for systems in which high-electron spin molecules (molecules with a large number of unpaired electrons) [29, 30] interact with metals of low density of states at the Fermi level or low nuclear mass. Besides that non-adiabatic effects have to be taken into account when the time scale of electron transfer of electrons, originating from electro negative molecules, becomes comparable to nuclear motion [10].

As an example for these processes one can regard the hydrogen recombination reaction on a metal (e.g. platinum) which is sketched in figure 2.1. The also occurring vibrational excitation of the metal substrate (heat - adiabatic process) is shown on the left side. However, this adiabatic picture holds only when one studies slow happening events (meaning a slow rearrangement of charges during bond formation and breaking), then the Born-Oppenheimer approximation (decoupling between the nuclear and electron motion) [31] remains valid. These events lead then to the sketched excitation of the substrate's lattice.

When the mentioned processes (rearrangement of charges during bond formation and breaking) are fast, deviations from the Born-Oppenheimer approximation occur. Then electronic excitations of the metal are created. This is summarized under the concept of non-adiabatic effects. Such electronic excitations that occur during the hydrogen recombination reaction are sketched on the right side of figure 2.1).

Or another example discussed is the deexcitation of highly excited NO molecules in terms of coupling of vibrational motion to electron–hole pairs, when they are scattered from surfaces [11, 32, 33].

2.1 Adiabatic or non-adiabatic reaction?

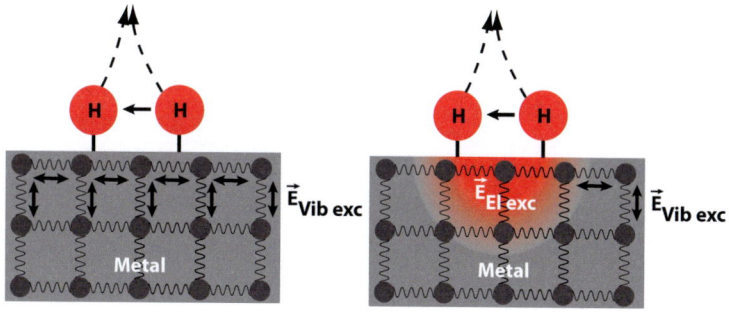

Figure 2.1: **Left side:** Vibrational excitation and **right side:** with additional electronic excitation during the hydrogen recombination reaction on a metal substrate being created.

These processes which happen with rapid and exothermic reactions on metal surfaces are underpinned by the Anderson orthogonality theorem [34]. This theorem states that the ground state of a many body Fermion system with infinite extension is orthogonal to the ground state of the same system in the presence of a localized scattering potential [35–38]. What means that an infinite Fermi system cannot remain in the ground state if a localized potential is suddenly switched on. This leads in metals to electron-hole pair excitations which are generated via a symmetry breaking process [37].

This symmetry breaking process will now be discussed in more detail [10, 37, 39]. Therefore we have a closer look on the bond formation of two species A and B and their corresponding potential energy surfaces in figure 2.2. Here their ionic ($A^+ + B^-$, V_{ion}) and covalent diabatic (A+B, V_{cov}) potential curve in dependence of their core distance are shown. For small core-core distances the ionic (V_{ion}) potential curve is favored, whereas for larger distances the covalent (V_{cov}) diabatic potential curve is energetically favored. At the crossing point of the two diabatic potential curves several opportunities (and transitions) are possible. When these diabatic potential curves are combined, one can get adiabatic potential energy curves, respectively surfaces (PES). Thereby the Born-Oppenheimer approximation [31] is taken into account, resulting in a neglect of the coupling between the core and the electron movement. The occurring diabatic crossings are denoted in figure 2.2 as a blue dotted line. The lower curve represents the ground state and the upper curve the excited state. The difference between these two states at the crossing point is denoted as V_{cross}. Assuming an adiabatic reaction scheme, the

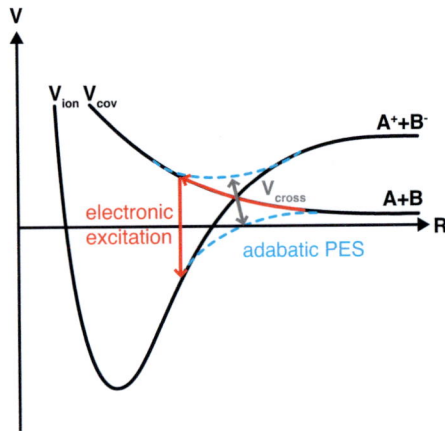

Figure 2.2: Ionic (V_{ion}, $A^+ + B^-$) and covalent diabatic (V_{cov}, A+B) potential curves plotted in dependence of their core distance. Possible adiabatic crossing (blue dotted line), the non-adiabatic crossing (red arrows) and the crossing point (grey arrow) are marked. Modified from [37].

bond formation would happen via the lower ground state curve. Excess energy would be released in excitation of rotational and vibrational modes; i.e. heat. A non-adiabatic pathway is marked with the red arrows. In this case the core movement could be so fast that the electrons would not be able to follow immediately (resulting in a deviation from the Born-Oppenheimer approximation). Due to the fact that the electronic system is not able to adopt rapidly enough the core-core approach, an excited state is formed in this non-adiabatic pathway. Meaning the whole system is past the crossing (follow the red arrow towards smaller core-core distances) in an excited state (A+B), while the energetically favored ground state is not occupied. The then following relapse of the excited electron into the ground state is marked with the second red arrow. This relapse process occurs with a certain probability. Smaller crossing differences V_{cross}, big velocities of the occurring reaction and a steep crossing lead to high probabilities for the electron relapsing process. However, in general these electron relapse probabilities per event are quite small ($\approx 10^{-5}$ per event) [10,11,37,40].

A detection of these kind of processes where excited states are populated, but the energy of the excited electrons is not sufficient that they can be emitted out of the surface, is not easy. Especially, because a detection of the hot electrons has to be done before they thermalize. Such thermalization is a conversion of the

2.1 Adiabatic or non-adiabatic reaction?

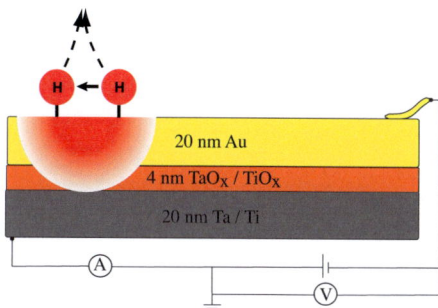

Figure 2.3: Schematic drawing of the hydrogen recombination reaction excitation process on the surface of a metal–insulator–metal sensor.

electron's excess energy into vibrations of the substrate, which happens whithin some 100 fs. The detection of such hot electrons as a current is therefore not easy. For a detectable current the excited electrons have to be spatially separated from their respective holes. This can be obtained via an internal potential barrier. This internal barrier then works as a kind of energy filter and separator for the excited carriers and their wide energy distribution. So when this internal barrier is different for excited electrons and holes, it can work as an energy filter of the carrier species. E.g. only electrons with sufficient high energy can overcome this barrier and are thus separated from their respective holes (when the barrier for holes is higher than for the electrons). Chemoelectronic sensors like **m**etal–**i**nsulator–**m**etal systems (MIM), Schottky diodes (metal–semiconductor sensor (MS)) or **m**etal–**i**nsulator–**s**emiconductor sensors (MIS) contain such an internal barrier (e.g. oxide band gap) and enable a detection of hot electrons without the limit that they have to be emitted out of the surface (exoemission) [4, 7, 8, 10, 12, 13, 41–45].

For example in a thin film MIM system an electron or a hole can be excited by the reaction taking place on the topmost surface (e.g. via the mentioned hydrogen recombination reaction), tunnel through the insulator and be detected as a net current of electrons or holes in the back metal electrode. However, this transport has to be finished within some 100 fs (see above), therefore one needs thin film sensors which lie in the nanometer thickness range (compare figure 2.3) to measure the excited carriers as a current. The tunnel barrier between the metals is a bit tunable via an applied bias voltage (± 1 V), what enables a kind of

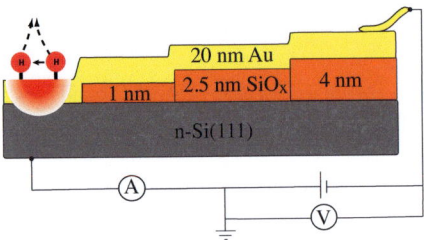

Figure 2.4: Schematic drawing of a stepped-MIS sensor with four different thick oxides and the hydrogen recombination reaction excitation process on the surface.

spectroscopy of the charge carriers [7, 9, 14, 46].

In a metal–semiconductor system a Schottky barrier is formed due to a singular charge transfer when a metal approaches a (e.g., n-doped) semiconductor. (By this charge transfer then the internal barrier is built up) [47, 48]. An electron has to overcome this barrier on the way from the metal to the semiconductor. A few of the excited electrons have enough energy to overcome this barrier. Hence, a part of the hot electrons, which are able to reach the potential barrier within their lifetime, can overcome this barrier and be detected as a so called chemicurrent in the semiconductor. With this kind of sensors one can only probe electrons or holes, according to the doping of the semiconductor. Additionally the doping determines the height of the Schottky barrier. [3, 46]. When MIS systems, which are a mixture of a MIM and a Schottky system, as used in this work, one can additionally apply bias voltages to modify the barrier height. Thus spectroscopy should be possible due to its insulating layer between the metal and the semiconductor. Such a stepped-MIS system is shown in figure 2.4 (here shown for differently thick oxides - what is the origin for the name stepped–MIS) and will be a working horse in this work.

2.2 Molecular beams

Molecular beams are focused beams of molecules, which fly under collision free conditions through a vacuum. Molecular beams are a valuable instrument to study the interaction of gases with surfaces in the field of physical chemistry. With such molecular beams one can direct gases to surfaces at low pressures, exhibiting a defined velocity and ambient direction of motion, even with a defined internal energy distribution. In that way surfaces can be exposed to a gas, without a significant effect on the ultra high vacuum (UHV), which is used for the study of the elemantary steps of reactions in physical chemistry. Hence from pulse to pulse only the number of occupied adsorption sites on the suface changes, in case that the molecules adsorb on the surface. Moreover it is possible to vary the kinetic and internal energy of the gas molecules by warming the nozzle, as well as by adding of heavier or lighter inert gases (seeding of the gas). Thus one can use the molecular beam technology to carry out many types of experiments - and study chemical reactions like catalysis or surface moderated reactions [49–64].

A molecular beam is generated by an expansion of a gas. In general one distinguishes two kinds of molecular beams: effusive beams and supersonic beams or nozzle beams. To describe these beams one uses the Knudsen number Kn, which reflects the relation between the mean free path length λ in the exhaust port and the nozzle diameter d [49, 64].

$$Kn = \frac{\lambda}{d} \tag{2.1}$$

With a Knudsen number greater than one ($Kn > 1$) one speaks about effusive beams. This is the case when the mean free path length of the effluent gas is greater than the nozzle opening. For example when the gas feeding pressure is very low, such Knudsen numbers can be derived. This entails, that between the effluent gas molecules few collisions occur. Their internal energy distribution is preserved as far as possible, instead they collide more often with the nozzle wall, which leads to an equilibration with the nozzle wall temperature. Therefore the particles have only a low thermal energy and a wide energy distribution and exit the exhaust port in a cosine like flux distribution (see figure 2.5) [64, 65]. Then by means of apertures a part of the beam can be selected. Generally speaking, effusive beams have less intensity due to their low feeding pressure. Effusive beams are mostly used to apply atoms or molecules homogeneously on surfaces.

If the mean free path λ of the gas particles is substantially smaller than the nozzle diameter, for example with several bar feeding pressure, the Knudsen num-

2 Basic information

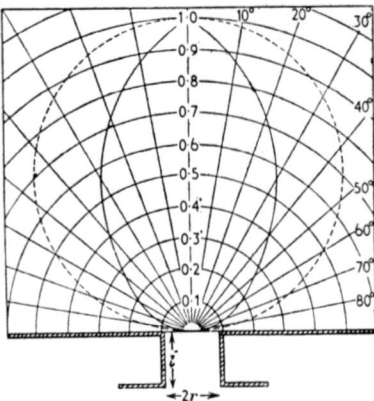

Figure 2.5: The angle distribution of particles effusing from a short circular channel. The straight line marks the case for equal dimensions of length and diameter of the channel, the skattered line marks the case for neglectable length. Taken from ref. [64].

ber is smaller than one ($Kn < 1$) and one speaks of supersonic beams or nozzle beams. In this kind of beams numerous collisions between the particles occur in the region of the nozzle. Through this, the at first undirected, accidental movement of the particles (internal energy like rotation and vibration) is converted into a hydrodynamically beam flow [65, 66]. Since in that way no heat exchange with the surroundings occurs, one speaks of an adiabatic beam expansion behind the nozzle [51]. On account of the expansion, the density of the gas decreases fast in beam direction, respectively the pressure and the number of the collisions, and the thermodynamically flux is converted into a free molecular flow. In this free molecular flow no collisions between the gas molecules occur. One speaks of a freezing of the, up to that time achieved, energy distribution regarding the internal degrees of freedom and translational energy. The point where this change occurs is labelled freeze in point and lies some millimetres behind the nozzle opening [49, 67].

By the prevailing different particle speeds (speed of the mass flux v) different zones of a supersonic expansion directly behind the nozzle can be characterised by the Mach number M (2.2).

$$M = \frac{v}{c} \qquad (2.2)$$

2.2 Molecular beams

And the local speed of sound c for an ideal gas is given by equation 2.3.

$$c = \sqrt{\gamma \cdot R \cdot \frac{T}{M_m}} \tag{2.3}$$

Here γ is the relation of the heat capacities (c_p/c_v), R the ideal gas constant, T the gas temperature and M_m the molar mass of the particles. One should note, that the big Mach numbers in the beam are not due to a high flux velocity, but on account of the small translation temperature T_{trans} (or gas temperature). With a Mach number M greater than one, one speaks of supersonic beams, with a small speed v of the mass flux one speaks of an effusive beam ($1 \geq M \geq 0$) [49].

From the point, where the pressure nearly reaches the background pressure of the chamber, the expansion breaks off in a shock wave. Therefore, the beam center is separated by a skimmer in a molecular beam apparatus and the regular movement of the beam is not disturbed by such a shock wave. The part of the nozzle beam, which is not flying in beam direction vertical to the nozzle opening after the expansion into the vacuum, is peeled off and one gets a fine directed beam. This formed beam is led via the other pumping stages and another aperture into the Ultra High Vacuum chamber. (So the skimmer also serves as a pressure stage that separates the next chamber from the recipient.)

Hence, the beam profile, just as the angular distribution of the particle beam, are a composition of two components when they exit the last pumping stage:

- The geometrical beam profile due to the nozzle and skimmer configuration and eventual connecting channels between the different pumping stages

- and a background distribution caused by effusive leaving particles of the last pumping stage.

The beam intensity is limited by the connection between the last two pumping stages [49, 68]. Hence, the pressure in the last pumping stage must be low enough not to disturb the measurements [51].

Assuming an ideal expansion, the translational energy of the gas particles in the beam can be calculated by the probability that particles reach the nozzle and the known heat capacity c_p at constant pressure and the equilibrium temperature T of the gas before the expansion by formula 2.4 [66, 70–72].

$$E_{\text{Trans}} = c_p \cdot T + \frac{1}{2} \cdot k_B \cdot T \tag{2.4}$$

A variation of the translation and internal energy is possible by the choice of the nozzle mode of operation. Depending on how the nozzle is operated, whether

2 Basic information

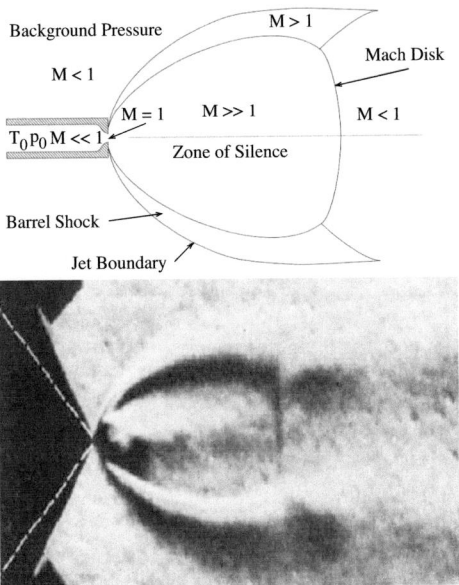

Figure 2.6: **Top:** Machdisk division of a nozzle expansion, based on [49] **below:** Schlieren photography of the interaction of the molecular beam with the background, taken from [69].

number of atoms	molecular geometry	c_p / k_B	example
1	spherically symmetric	3/2	He, Ar
2	linear	7/2	H_2, O_2
3	linear	13/2	CO_2
3	angular	6	H_2O

Table 2.1: Examples of the heat capacity at constant pressure for different gases.

the nozzle is heated or whether to the reaction gas heavier or lighter inert gases are added (seeding), beam energies between the thermal energy (≈ 40 meV) and approximately 20 eV are possible to adjust [63, 66, 73]. Since the dissociation energies of chemical bonds and the activation energies lie in this area, supersonic beams are often used to study chemical reactions [52–58, 60, 61].

In the course of the time, procedures were also developed to use the molecular beam not only in a continuous, but also in a pulsed operating mode. The formation of quasi pulsed beams is possbile in an easy way by using a rotating disc with an opening as a chopper which interrupts the continuous beam [70, 74].

The first principle used for the formation of proper pulsed molecular beams was the current loop mechanism of Dimov in 1968 [75], which was refined by Gentry [76, 77]. Injection nozzles similar to the ones in a car are used nowadays often. The nozzle opening is closed by a moveable teflon popped. This teflon popped is then pulled back electromagnetically to open the nozzle and a flow of gas can occur [71]. By controlling the electromagnetic pulses a pulse variation of the molecular beam is possible [51]. However, there are two main problems with pulsed molecular beams [51]. The first problem deals with the dynamics of the expansion and, hence, is tied up with the question, which pulse lengths are needed, to form quasi steady flux conditions after opening the nozzle. (Here ideal conditions regarding an opening in a rectangular shape are assumed.) The required time to enable steady flux conditions is generally smaller than some microseconds, and, hence, only with extremely short pulses really decisively. The second problem deals with the not ideal behaviour of the valve gear during the opening and closing procedure. At the beginning and at the end of the pulse, when the maximum retraction of the valve is not reached yet, the expansion is hindered by the teflon popped and the formation of a quasi stationary flux is delayed [51]. The exact dependence of this delay time, limited by the valve mechanism, was studied by the rotationally and vibrationally relaxation of the molecules which expanded with a carrier gas as a function of the time and the distance from the nozzle [78].

Therefore molecular beams can be used to direct a beam of molecules onto a surface. Even a pulsed dosage and different beam energies are adjustable.

2.3 Cyclic Voltammetry

In cyclic voltammetry the electrode current is measured as a function of the applied voltage to the electrochemical cell. The nowadays-used cyclic voltammetry is an extension of the simplest potential-time profiles, which are used in the so-labelled linear sweep voltammetry. In this technique the potential is swept from an initial value $E_{initial}$ to a final potential E_{final} and the responding current of the electrochemical cell is recorded. In contrast to this linear sweep voltammetry in the cyclic voltammetry technique the potential is reversed when it reaches the potential E_{final}. After reaching this value E_{final} the potential is scanned back to the initial potential $E_{initial}$. Then there are several possibilities, the potential sweep may be halted, again reversed or alternatively continued further to a value. Thus a kind of triangular potential cycle is obtained as shown in the top graph of figure 2.7 [79, 80].

The sweep rates used in the most common experiments range from a few mV/s up to a few hundred mV/s. The resulting electrode current is then measured as a function of this applied voltage to the electrochemical cell. Commonly several peaks will be observed there. From observing how these peaks appear and disappear as the potential limits and sweep rates are varied, one can determine how the processes, represented by the peaks, are related. There is even the chance to determine the role of adsorption, diffusion, and coupled homogeneous chemical reactions by recording the sweep rate dependence of the peak amplitudes. The most worthwhile information can be gained by the difference between the first and subsequent cyclic voltammograms. But a kinetic study can only be accurately obtained from an analysis of the first sweep [79, 80].

For obtaining these results of an electrochemical study a three electrode set-up is generally required. The set-up consists of the reference (RE), a counter (CE) and a working electrode (WE). The working electrode (platinum, nickel, gold etc.) is where the electrochemical reaction takes place. As counter electrode it is most common to use a platinum mesh, so that a large surface area is available to allow a substantial current to flow, otherwise platinum wires are used. The reference electrode is used for maintaining the potential drop across the electrode – electrolyte interface. In general the reference electrode is a saturated calomel electrode (SCE) or a saturated silver – silver chlorine electrode (Ag/AgCl). The applied potential between the working electrode and the reference electrode can be described by the following equation, where $\phi_s - \phi_l$ is the potential drop across the working electrode-electrolyte interface, iR is the poten-

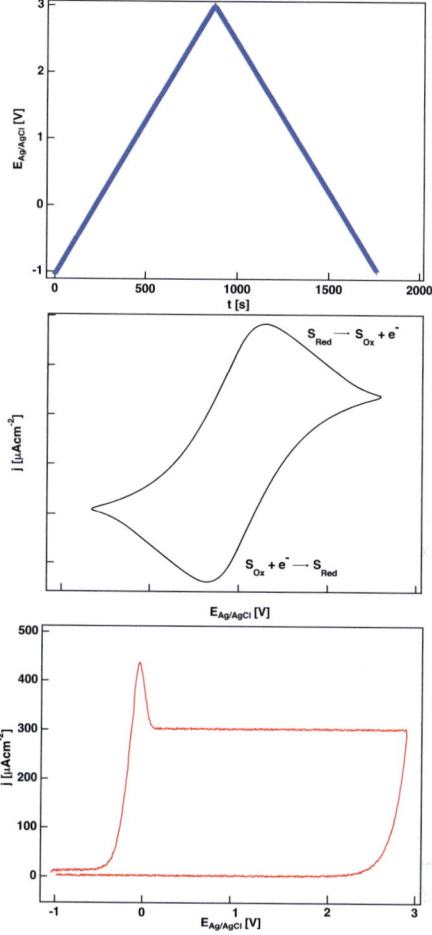

Figure 2.7: **Top graph:** Triangular potential cycle. **Middle graph:** Cyclovoltammogram for a reversible oxidation and reduction process, compare [79]. **Bottom graph:** Cyclovoltammogram for an irreversible oxidation, e. g. aluminum, tantalum or titanium oxidation [81–84].

tial drop that is caused by the current i which flows across the electrolyte with the resistance R: $E = (\phi_s - \phi_l) + iR + (\phi_l - \phi_{ref})$. The difference of $\phi_l - \phi_{ref}$ is the potential drop across the solution-reference electrode interface [79, 80]. For controlling the potential between the working electrode and the reference electrode at a fixed and selected potential, and to measure the current that flows between the working and the counter electrode, a potentiostat is used.

In case of an reversible oxidation and reduction process, one would get a symmetric cyclovoltammogram as displayed in the middle graph of figure 2.7. The oxidation process releases electrons who cause a higher observed current. For the reduction process electrons are needed and a lower current can be observed, which is equal regarding its absolute value to the oxidation current in case of a reversible reaction.

Asymmetric cyclovoltammograms, as shown in the bottom graph of figure 2.7 can be obtained when the electrochemical oxidation is irreversible, what is the case e. g. for aluminum, tantalum and titanium [81–84]. A constant oxidation current plateau is followed after the small overshoot peak at the beginning of the oxidation. Oxidation of such thin films can be carried out with the help of a droplet cell [85, 86], shown in figure 2.8. Most commonly an acetate buffer electrolyte is used to minimize parallel corrosion processes during oxidation [87–89].

One specialty in the oxidation of semiconductors should be noted here, since one has to bear in mind the dopant type (n- oder p-Si) during oxidation. For p-type silicon the anodic reaction kinetics are not affected by an illumination since the reaction consumes holes, which are the majority carriers and therefore their concentration is only little affected by illumination. However, for n-type silicon, as it is used in this work, one has to be aware of sufficient illumination during the oxidation process. Since in this oxidation case the interface is reversely biased and either holes have to be generated or electrons have to be injected from the electrolyte to sustain the oxidation reaction. Therefore one would have to apply an extra voltage (in comparison to an anodization of p-Si) to drive the current, when the oxidation is performed [90]. Thus always care was taken to provide sufficient illumination during the anodic oxidation.

Further details regarding the electrochemical oxidation processes will be given in the next chapters where the results are discussed.

2.3 Cyclic Voltammetry

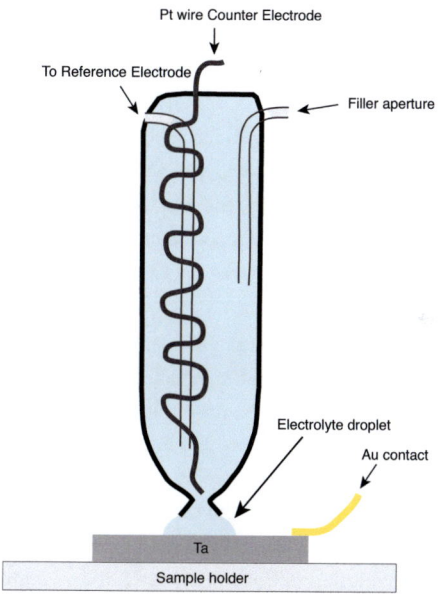

Figure 2.8: Schematic drawing of an electrochemical droplet cell set-up.

2.4 Water formation reaction

In 1823 Döbereiner informed his minister J. W. Goethe about his observation that hydrogen reacts with oxygen in presence of platinum at ambient temperature in pure contact with water [91, 92]. This news, together with similar observations in other laboratories, led Berzelius to introduce the idea of catalysis for this mysterious class of chemical phenomena [93].

In consideration of the fundamental importance, this reaction was the object of many studies. Despite the ostensible simplicity the exact reaction mechanism remained unexplained. Among other things this oxidation of hydrogen was studied due tue its model character for a heterogeneous reaction by many researchers as Kasemo and Rosén [94–96], Ertl [93, 97, 98], Verheij [99–101] and many others on Pt(111). Ertl even was awarded with the Nobel prize in chemistry 2007 for his studies of chemical processes on solid surfaces (in particular he was awarded for the continous studies regarding e.g. i) the mechanism of the catalytic reaction in the Haber-Bosch process, where nitrogen molecules react with hydrogen molecules to form ammonia and ii) reactions like the CO oxidation or oxygen adsorption on platinum) [1].

In this oxidation of hydrogen the catalyst is quite efficient, because the reaction already runs off at 120 K. Moreover, the structure of Pt(111) is very well understood and can be obtained by different preparation steps. Besides the interaction of this surface with both reactands (oxygen and hydrogen), as well as the reaction with the product water was intensively studied [99]. Therefore one would choose the water formation as one of the first molecular reactions to study with the chemoelectronic sensors. Maybe the underlying reason for the high efficiency of the catalyst in the oxidation of hydrogen, even at low temperatures as 120 K, can be found. Thus one would consider studies especially in the low temperature regime, but for a complete picture as well in and above the temperature range where the product water desorbs with a reasonable efficiency.

In the following an overview of the models regarding the oxidation of hydrogen on platinum is given. Verheij and his coworkers [99, 100] prefer a reactive site model on account of their studies of the water formation reaction in a temperature range from 100 to 1400 K, where depending on the temperature regime different effects show up [100, 102]. The reactive sites are labelled in the following *r.s.*. In this reactive site model one assumes that only a few special platinum sites are catalytic active. Accordingly the reaction does not run homogeneously at the surface. In contrast to that other models suppose that all surface sites are catalytic

2.4 Water formation reaction

active or can become catalytic active [93, 94, 96–98].

In the following the reactive site model of Verheij [99] is introduced for cases where the coverage of the platinum surface with oxygen is small (< 0.05 monolayers). Afterwards it is described, how Verheij [100] expanded this model, based on experiments above 300 K, by light modifications to the low temperature regime (below the desorption temperature $T_{desorption}$ of water on Pt(111), meaning below 170 K [98]). In this model the adsorption on and desorption (of the reactands H_2 and O_2) from the surface are considered as processes occurring with equal reaction rates all over the surface. For the reaction steps in which H_2O is formed it is assumed that they occur at reactive sites r.s.. Moreover species which are in the neighborhood to an occupied reaction site are different from other species. These neighborhood sites are labelled with $(_*)$. Adsorbed atomic oxygen at such a place is labelled as O_*. Just analogue to the homogeneous reaction models, in this model the different ways from the intermediate product OH to the product H_2O are considered. Successive addition of hydrogen and an alternative reaction sequence via two oxygen atoms, in analogy to a disproportionation, are described under the assumption of reactive sites and transport of the slowly diffusing O atoms and the OH molecules.

In the following reaction scheme, the species adsorbed on the surface are not explicitly marked; instead the species not adsorbed on the surface like $O_2{}^{Gas}$ are explicitly marked.

$$O_2{}^{Gas} \rightleftarrows 2\,O \qquad (2.5)$$

$$H_2{}^{Gas} \rightleftarrows 2\,H \qquad (2.6)$$

$$O \rightleftarrows O_{r.s.} \quad \text{(diffusion)} \qquad (2.7)$$

$$H + O_{r.s.} \rightleftarrows OH_{r.s.} \qquad (2.8)$$

$$OH_{r.s.} \rightleftarrows OH \quad \text{(diffusion)} \qquad (2.9)$$

$$OH_{r.s.} + H \rightleftarrows H_2O \quad \text{(sequential addition of H)} \qquad (2.10)$$

Then Verheij and Hugenschmidt [99] suggested a reaction sequence where they introduce the species H_2O_2 to make this model applicable to other surfaces. This species could be relatively stable or an activated complex which dissociates almost immediately after its formation.

$$O \rightleftarrows O_* \quad (\text{diffusion}) \tag{2.11}$$

$$OH_{r.s.} + O_* + H \rightleftarrows H_2O_{2\,r.s.} \tag{2.12}$$

$$OH \rightleftarrows OH_* \quad (\text{diffusion}) \tag{2.13}$$

$$OH_{r.s.} + OH_* \rightleftarrows H_2O_{2\,r.s.} \tag{2.14}$$

$$H_2O_{2\,r.s.} \rightarrow H_2O_2 \tag{2.15}$$

$$H_2O_2 \rightleftarrows H_2O + O \tag{2.16}$$

The second for the formation of the intermediate product H_2O_2 necessary oxygen atom diffuses to the reactive site either after step (2.11) or in step (2.13) as a part from OH. After the intermediate product H_2O_2 was formed via step (2.12) or step (2.14), it can break down via step (2.15) and step (2.16) finally to H_2O. This formed H_2O, via the steps with the H_2O_2 (2.16), or via a sequential addition of H (2.10), can desorb from the surface (2.17). (Since the model is made for temperatures $T > T_{\text{desorption}}$ where the desorption rate is high enough that desorbing gas particles can be detected).

$$H_2O \rightarrow H_2O^{\text{Gas}} \tag{2.17}$$

This reactive site model, which was based on experiments above 300 K, was expanded by Verheij [100] to the low temperature range by slight modifications. The first step in the formation of H_2O via the mechanism mentioned above at reactive sites in the low temperature regime (below 170 K) is, after the dissociation of the reaction partners (2.5), (2.6) and diffusion of an oxygen atom to a reactive site (2.7), the reaction of a hydrogen atom with an oxygen atom $O_{r.s.}$ at a reactive site (2.8), just as in the model for the high temperature range. However, this $OH_{r.s.}$ formation reaction is assumed to be irreversible at low temperatures. Bit by bit the $OH_{r.s.}$ reacts with a second H atom, either by itself under formation of $H_2O_{r.s.}$ which reacts then with a neighbouring O atom; or however with a neighbouring O atom. Both processes lead to a H_2O_2 or two OH intermediate products at the reactive site.

$$OH_{r.s.} + H + O_{r.s.} \rightarrow \text{intermediate product(s)} \qquad (2.18)$$

This reaction can run only until the reactive sites are completely surrounded by intermediate products. At that time H_2O can be formed; a front between the surrounding intermediate products and the rest of the (oxygen covered) surface is formed. Before the next reaction can run off, a H atom must leave the reactive site by a process in which it remains bonded to an O atom. If now one H atom reaches an intermediate product at the reaction front by such a transport process, a H_2O molecule is formed, which has an O atom as a neighbour. Then this H_2O molecule can react with the oxygen and shift the reaction front by one reaction site. For these reaction steps the equations (2.19), (2.20), (2.21), (2.22) are given, in which H_r is hydrogen that is adsorbed at the surface together with H_2O and the intermediate products.

Besides, it is supposed that the O atoms are quite immobile, that OH is mobile only in reaction (2.19), where it must reach $OH_{r.s.}$ to form H_2O_2. Additionally one thinks that the mobility of H_2O and H_2O_2 is limited by their surrounding neighbours.

$$OH_{r.s.} + OH \rightarrow (H_2O_2)_{r.s.} \qquad (2.19)$$

$$(H_2O_2)_{r.s.} + H_r \rightleftarrows OH_{r.s.} + H_2O \qquad (2.20)$$

$$H_2O_2 + H_r \rightleftarrows OH + H_2O \quad \text{(transport)} \qquad (2.21)$$

$$H_2O + O \rightarrow H_2O_2 \quad \text{(front movement)} \qquad (2.22)$$

Reaction (2.22) leads to an increase of the coverage with intermediate products. It is the reaction in which the real intermediate products are formed. Verheij [100] supposes that this reaction runs irreversible below 170 K. Reaction (2.19) is the only reaction step that lets the covering with H_2O and H_2O_2 rise. As H_2O_2 can be transformed by means of reactions (2.20), (2.21) into H_2O, one can name reaction (2.19) as the actual water formation step.

If for example a surface with reactive sites, which is covered with H and O atoms at 100 K, is heated from 100 K, water is formed when the temperature rises

2 Basic information

above 120 K. But if the heating rate is too high, the reaction will not be completed, before the H_2O desorption temperature range of approximately 170 to 220 K [98, 100] is passed. The desorption of H_2O which surrounds the reactive sites makes a transport of reacted hydrogen to oxygen on the terrace sites impossible, so that these O atoms have to diffuse to a reactive site to react to H_2O. Because the O diffusion is rather slow, a part of the hydrogen is able to desorb, before a reaction can take place, meaning atomic oxygen can remain on the surface [100].

Moreover, with this model the dependence of the reactivity on the surface properties is explainable, because the water formation reaction proceeds only on active sites. If the concentration of reactive sites is low (low reactivity), oxygen atoms have to diffuse a longer distance before they can react. This causes an increase of the diffusion times. If the diffusion time is of the same dimension as the duration of the experiment, it implicates that the distribution of the oxygen atoms on the surface is inhomogeneous and depends on the covering at the beginning of the experiment. Because the H_2 sticking probability does not vary linearly with the covering of oxygen, one can expect that also the water formation rate depends on the starting conditions. Hence with a surface with less reactive site, less water is formed after this model.

But the roles of reactive sites, as suggested by Verheij [100], is not consistent with the data of Ertl and his coworkers [93, 97, 98]. They favour a mechanism for the water formation reaction via OH without the formation of H_2O_2.

After the dissociation of the reaction partners they formulate as well a reaction to OH (compare equation 2.8), but without requiring a reactive site. Afterwards the formed OH reacts with H to H_2O (see reaction equation 2.10 without the need of an active site), which then reacts to OH. This reaction (2.25) consists of two single steps where in each case a H_2O molecule is consumed (2.23), (2.24).

$$H_2O + O \rightarrow 2\,OH \qquad (2.23)$$

$$H_2O \rightarrow OH + H \qquad (2.24)$$

$$2\,H_2O + O \rightarrow 3\,OH + H \qquad (2.25)$$

This sequence is autocatalytic. Since if two OH molecules form two H_2O molecules in reaction step (2.10), step (2.25) leads to a formation of three OH molecules. This autocatalytic acceleration mechanism is consistent with the observeable induction period of the formation of water at low temperatures [103].

Also one can see from their scanning tunnelling microscope (STM) data that the oxygen layer is converted completely into OH before water is formed. H_2O is only stable if all neighbouring O atoms were converted into OH. This leads one to expect an average induction period for the water formation reaction at deep temperatures if the site specific rates are integrated about the whole surface. Whereas the reaction step (2.25) can be understood at low temperatures (even at 110 K) as a quick reaction step, the reaction rates decrease for the H_2O production at temperatures above 170 K [97]. One connects this with the desorption temperature of H_2O of 170 K [98] (Verheij mentions the above quoted desorption temperature range of 170 to 220 K [100]), because H_2O is required to connect both steps. When the water formation is carried out below this temperature range, e.g. at 120 K, the water formation reaction terminates when a complete covering with H_2O of the surface is reached. Moreover a pre-adsorption of H_2O, which takes part in the autocatalytic step, reduces the induction period for the H_2O formation [98].

But the just described reaction mechanisms are no longer efficient if not enough H_2O remains on the surface. Because reaction (2.25) needs water to generate, together with reaction (2.10), the autocatalytic process. Under ambient UHV conditions this critical water covering is only scarcely reached with the water desorption temperature of 170 K. Above this temperature water can be formed only by the successive H addition reactions (2.8) and (2.10) (without consideration of the reactive sites). Therefore a change in the reaction rate should be observable above the desorption temperature of water. Indeed Völkening et al. [93] were not able to observe a significant reaction at 200 K. On the basis of their STM images they observe at the beginning of the dosing from H_2 a disappearance of some O atoms. Then the reaction stops, even with long H_2 dosing times no other reaction can be observed.

On account of a similar behavior between 220 and 250 K [102] they consider that the reaction (2.8) is a slow step and limits the reaction rate at higher temperatures. Additionally they consider that the steps (2.10) and (2.25) have to be quick.

Hence Völkening et al. [93], conclude that the mechanism of the catalytic oxidation of hydrogen on platinum is determined crucially by the presence of the reaction product water on the surface. This mechanism is therefore a function of temperature and pressure. In presence of enough H_2O at the surface these species act autocatalytically, whereas with small concentrations of water usual kinetics with successive steps applies.

3 Experimental design and sample preparation

In this chapter some basic information regarding the experimental design and the sample preparation is given. For the respective experiments carried out in this work, more detailed information can be found in the following "Measurements, Results and Discussion" section.

3.1 The Ultra High Vacuum recipient

The recipient where most of the Ultra High Vacuum (UHV) studies are carried out is shown in figure 3.1.

The chamber is equipped with a rotary pump, a turbo molecular pump and a titanium sublimation pump to generate a sufficient UHV of $5 \cdot 10^{-10}$ mbar. The titanium sublimation pump, which evaporates from time to time a new titanium film onto a liquid nitrogen cooled wall (to bind the adsorbed particles at that wall), is switched off some time before the measurements. Thus any disturbances are avoided.

One reaches a base pressure of around $5 \cdot 10^{-10}$ mbar when the chamber, with its pumping system, is heated for a period of about 72 hours. An ionisation manometer serves as a pressure gauge. The sample carrier is mounted on a manipulator which allows a 360° rotation around the z-axis and a movement in x-, y-, and z-direction. The sample is glued to the sample holder via a glass slide with silver conductive varnish. The sample can be heated with a tungsten filament at the manipulator. Cooling is possible due to a direct contact with the cold head of a helium cryostatic temperature regulator. The temperature of the sample can be measured with a resistance thermometer (Pt1000). Besides, the chamber is equipped with an electron beam evaporator, a quadrupol mass spectrometer and an atomic hydrogen source. In this atomic hydrogen source molecular hydrogen flows through a radiatively heated (tungsten wire as heater) capillary. That way the molecular hydrogen can be dissociated thermally to atomic hydro-

3 Experimental design and sample preparation

Figure 3.1: Photo of the used apparatus. In this photo one can see the molecular beam on the right side and the UHV recipient in the middle (marked with a red rectangle).

gen [42, 104, 105]. This atomic hydrogen can be directed to the sample and interrupted by a mechanically rotating disc shutter. Additionally this source can be operated in such a way that just excited molecules leave this source. This can be performed by passing a smaller heating current through the tungsten wire.

3.2 The molecular beam apparatus

A three stage supersonic molecular beam apparatus with a pulsed nozzle is used in this work. This nozzle (general valve, type 9), controlled by a nozzle pulse generator, built at the 'Forschungszentrum Jülich', has an opening diameter of 500 μm.

In its static condition it is closed by a teflon poppet, which can be electromagnetically pulled back, what opens the exhaust port and produces a gas pulse (as described above). The teflon poppet (the nozzle) in the first pumping stage of the molecular beam apparaturs is separated by the first skimmer from the second pumping stage [71]. In the first pumping stage an oil diffusion pump maintains a pressure of 10^{-4} mbar. The pressure in the third stage is lowered with the help of turbomolecular pumps to 10^{-7} mbar.

The passage of the molecular beam is enabled by an aperture (diameter 3 mm) between the third pumping stage and the main Ultra High Vacuum recipient. Additionally the whole molecular beam apparatus is separated by a plate valve from the main UHV chamber to avoid any disturbances when it is not operating. The distance between the nozzle and the first skimmer (diameter of 3 mm) cone can be adjusted and is set to 5 mm after an optimization procecure described in my diploma theses [106].

In figure 3.2 a photo of the nozzle skimmer arrangement in the molecular beam chamber and a photo of the nozzle under ambient conditions are shown as well as a schematic drawing of the molecular beam and its passage to the UHV recipient.

3.3 The ion beam source

Ion beams can be used for different applications such as sputtering (ablating a target like a sandblast cleaning) or ion beam etching and for ion beam analysis. The ions can be generated by different ways. During the interaction, ions loose energy due to nuclear and electronic stopping, de-excitation via Auger transitions, plasmon excitation, and photon emission [14, 107–114].

3 Experimental design and sample preparation

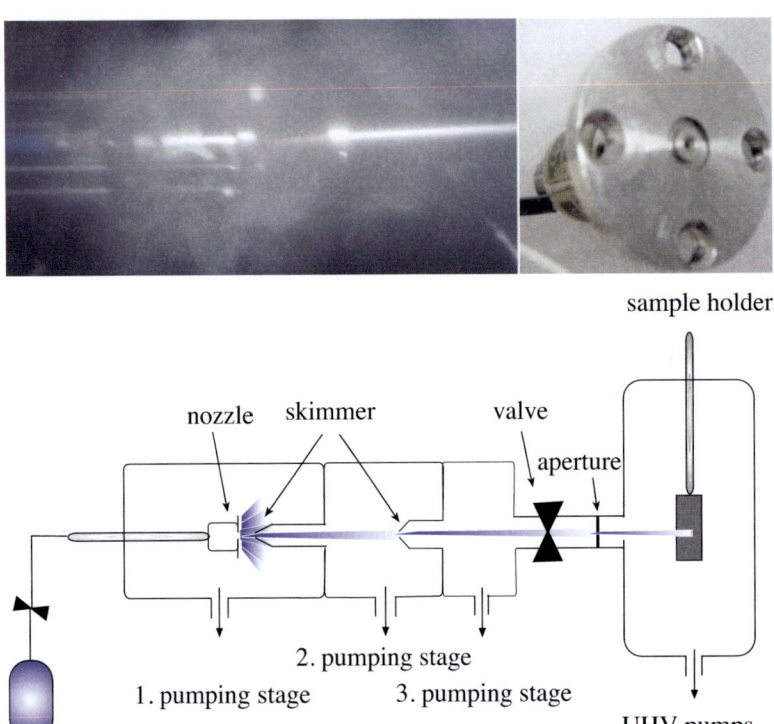

Figure 3.2: **Top:** Photo of the nozzle skimmer arrangement in the molecular beam chamber and a photo of the nozzle under ambient conditions. **Bottom:** Schematic drawing of the molecular beam and its passage to the UHV recipient.

3.4 Tunnel detectors

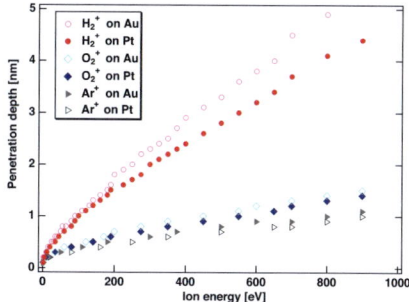

Figure 3.3: Simulated penetration depth in dependence of the ion energy for hydrogen, oxygen and argon ion impacting on gold and platinum surfaces [115–117].

In this work an Oxford applied research low energy ion source LIon 50 is used. It is operated with a heated filament and desired gas feed. Ion beams with energies up to 900 eV can be formed. Focus and deflection is possible via different voltages on apertures and lenses.

In figure 3.3 the penetration depth reachable with the ion source, gases and substrates used in this work is plotted in dependence of ion energies. These data were obtained via a simulation with the computer program SRIM where the mass 2 was used for molecular hydrogen and mass 32 for molecular oxygen [115–117].

3.4 Tunnel detectors

3.4.1 Metal–insulator–metal tunnel detector (MIM)

Metal–insulator–metal tunnel detectors consist of an e.g. 20 nm thick back metal electrode stripe (e.g. aluminum, tantalum, titanium) which is evaporated in an Ultra High Vacuum chamber, utilizing an electron-**beam** evaporator (tantalum, titanium) or a thermal evaporator (aluminum) on an isopropanol and ethanol cleaned glass slide. Then a part of the metal layer is electrochemically oxidized by means of cyclovoltammetry (see above). In such a way an e.g. 3 nm thick metal oxide layer is formed on the base metal. Then the top metal electrode (thermal: silver and gold 20 nm thick, e-beam: platinum 7 nm thick) is evaporated through a mask in crosswise arrangement to the back metal stripe. Silver conductive vanish is used to contact the bottom and the top electrode. A schematic

3 Experimental design and sample preparation

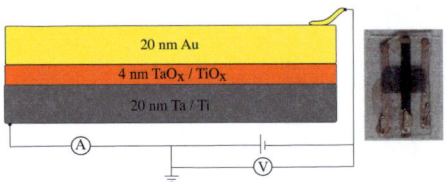

Figure 3.4: Schematic drawing and a photo of a metal–insulator–metal sensor.

drawing and a photo of a MIM sensor can be seen in figure 3.4.

3.4.2 The new stepped metal–insulator–semiconductor tunnel detector (stepped–MIS)

Deduced from Schottky diodes [118–120] and MIM-sensors a new class of chemo-electronic sensors was developed. This new metal–insulator–semiconductor (MIS) sensor, consists of a 525 μm thick silicon wafer (n-doped, $\rho = 7.5\,\Omega \cdot cm$), which is cleaned with ethanol, hydrofluid acid and water, an electrochemically formed silicon oxide layer and a 20 nm (or 7 nm e-beam evaporated platinum film), 4 mm wide gold stripe, which is evaporated in the Ultra High Vacuum by thermal evaporation through a mask on the sensor. The top gold (platinum) layer is contacted with a gold contact and silver conductive vanish. The silicon is contacted via a backcontact with silver conductive vanish. The preparation of the oxide via cyclovoltammetry allows one to prepare several sensors with different oxide thicknesses on one wafer, just by choosing the oxidation potential according to the film formation factor of 0.4 – 0.7 nm/V [121]) (e.g. 1 – 4 nm). In figure 3.5 a photo and a schematic drawing of a stepped Si–SiOx–Au (stepped–MIS) sensor and its contacts in a measurement set-up are shown.

3.4 Tunnel detectors

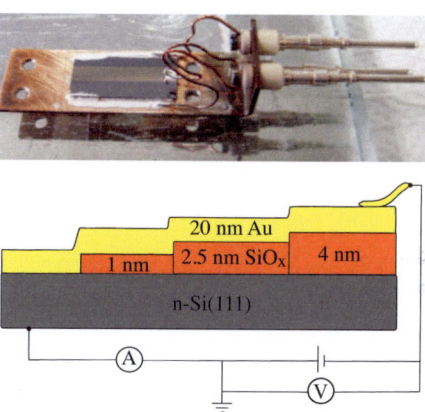

Figure 3.5: Photo and a schematic drawing of a stepped Si–SiOx–Au (stepped–MIS) sensor and its contacts in a measurement set-up.

4 Measurements, results and discussion

4.1 Molecular beam

The molecular beam section will present some characteristics of the molecular beam.

4.1.1 Translational energy of the particles

The translational energy (E_{Trans}) of the particles of the molecular beam can be derived via the particle velocity v_0 via $E_{Trans} = \frac{1}{2} \cdot m \cdot v_0^2$. Therefore the velocity v_0 of the molecular beam is measured with the **q**uadrupole **m**ass **s**pectrometer (QMS) (external readout, 2.4 kV, multiplier gain E3) mounted on a z-stage. Then gas pulses (compare figure 4.1) are recorded with the QMS at different positions. The maxima of the QMS-signal occur later in time with larger distance between the molecular beam and the QMS. Then one obtains the particle velocity v_0 via the time difference (Δ t) and the difference in the distance s ($\Delta s = 10$ cm).

This velocity can be changed when the flowing gas particles are heated by the molecular beam nozzle. Such a heatable molecular beam nozzle is sketched in figure 4.2. It was configured in allusion to a previous used nozzle [67, 70]. It consists of a macor plate which is mounted in front of the nozzle to hold the other equipment. This equipment consists of an Al_2O_3 tube of 2.5 cm length, which is surrounded by a coated constantan wire on a length of 1.5 cm and covered with a steel tube. When a current flows through the constantan wire, the Al_2O_3 tube is heated and so are the gas particles, coming from the gas feed through the nozzle into the Al_2O_3 tube. The temperature of the Al_2O_3 tube is measured with a Ni-NiCr thermocouple at the top position, separated from the heating parts by a ceramic ring. This measured temperature is assumed to be the real temperature of the molecular beam particles.

In that way a temperature dependent velocity and energy study of the molec-

4 Measurements, results and discussion

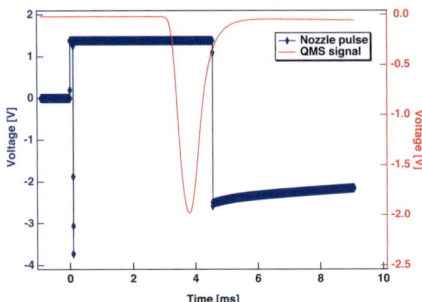

Figure 4.1: Example of a CO gas pulse (1.5 bar nozzle high pressure feeding pressure) signal measured with the QMS (red curve) and the nozzle control signal (blue curve) which opens the nozzle for a 3 ms gas flow of CO.

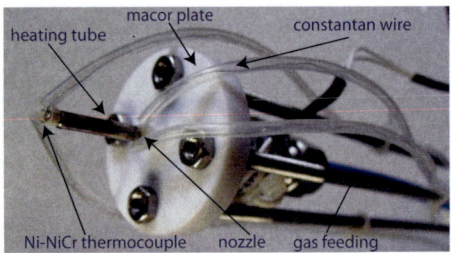

Figure 4.2: Photo of the heatable molecular beam nozzle. An Al_2O_3 tube of 2.5 cm is surrounded by a constantan wire on a length of 1.5 cm and covered with a steel tube. The other equipment parts are denoted. The nozzle is just behind the tube in the middle of the macor plate.

ular beam is possible. More information about how many particles can be dosed with these translational energies will be given in the next subsection.

4 Measurements, results and discussion

Figure 4.3: Photo of a 12 nm thick platinum film (2 cm · 1 cm), contacted with copper contacts and silver vanish, on glass mounted on the sample holder.

4.1.2 Particle numbers of the molecular beam

To determine the particle numbers of the molecular beam the established method temperature programmed desorption (TPD) is used. In TPD particles are adsorbed on a sample surface at a certain temperature, e.g. with a molecular beam. Then the sample is warmed up with a constant heating rate what drives particles to desorb from the surface with a reasonable probability ($T_{desorption}$). These desorbing particles can be monitored then with a mass spectrometer [122–124].

First one would think of a direct heating of an e.g. platinum film on glass slides, as shown in figure 4.3. In this direct heating method one uses the resistivity of the platinum film to heat the film using a power supply. However, some problems regarding the stability of the platinum film occur. One does not get reproducible heating rates when the samples are modified as shown in figure 4.4. Here Scanning Electron Microscope (SEM) images are shown of a 12 nm thick platinum film, which was heated and showed hot spots which modified the platinum film. These modifications can be clearly seen in these SEM images. As the upper SEM image shows a bubble like structure (probably due to the hot spots). Even a rip seems to occur, as can be seen on the right side of the lower SEM image. Furthermore, in the middle of this image one can see some blended like structures. Therefore one could think of electromigration effects, since currents of $\approx 5 \cdot 10^5 \text{A} \cdot \text{cm}^{-2}$ where applied [125], which are close to the electromigration limit of the platinum film of $10^5 - 10^6 \text{A} \cdot \text{cm}^{-2}$ [126]. But a definite interpretation of the mechanism of the formation of the bubble like structures goes beyond the scope of this work.

A better way is to heat the platinum film indirectly by direct heating of an underlying silicon substrate. Therefore a 10 nm thick, 1.8 cm long and 0.9 cm wide platinum film is evaporated on a hydrofluoric acid cleaned silicon n-Si(111) piece (2 cm · 1 cm). To enable this kind of heating, a new sample holder and manipulator was set-up and configured on the basis of existing manipulators, of different size, in the group [67, 70].

4.1 Molecular beam

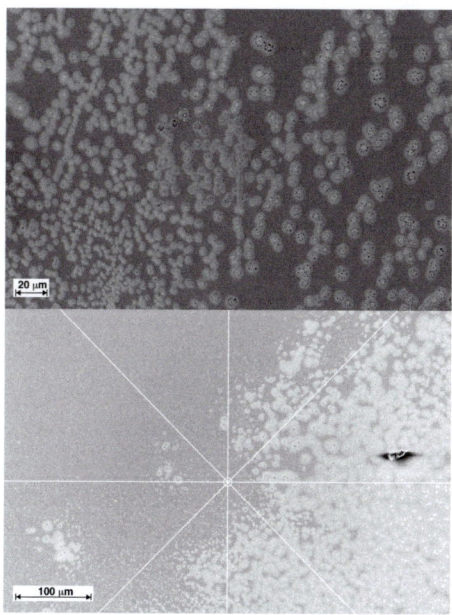

Figure 4.4: SEM images of a platinum film which was heated up to 560 K for 5 min. **Upper SEM image** shows bubble like structures of hot spots. **Lower SEM image** shows a rip and migration like structures of the hot spot bubbles.

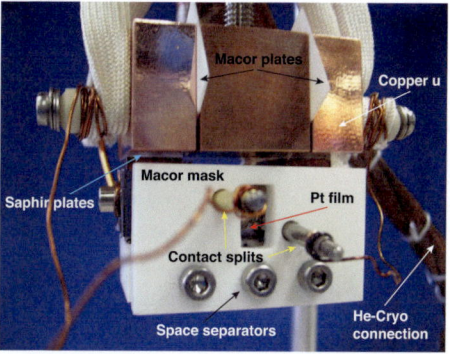

Figure 4.5: Photo of the new sample holder for direct heating of the samples, e.g. a platinum film on top of a silicon heat substrate. Different markings denote the special parts of the sample holder and its equipment.

4 Measurements, results and discussion

With this new sample holder sketched in figure 4.5 heating of the samples is performed by holding the e.g. silicon substrate, with platinum on top, with tantalum clamps and applying a current that heats the silicon sample due to its resistivity and thus the platinum film as well. Cooling is performed via a back connection to a Helium-cryostat. This back connection is mounted on the back side of the U shaped upper copper part of the sample holder. This copper U part of the sample holder is insulated in terms of temperature and electric conductivity via macor plates from the manipulator rod and its base (see wing like macor plates separating the copper U from the cube that is connected to the manipulator rod in figure 4.5).

The temperature is controlled using a Ni–NiCr thermocouple, which is mounted directly on the back side of the sample. Its temperature readout is calibrated each time for the sample (since the front e.g. platinum side temperature is the temperature that one wants to know and not the 'back silicon heat substrate' temperature) using a pyro meter. To ensure a good heat flux without electric conductivity of the bottom parts of the sample holder, saphire plates are used (see marking in figure 4.5).

To make front contacts on the samples possible, a special mask was designed which holds spring bolt contact splits which connect the sample with modest pressure. This white macor mask with the golden spring bolds contacting a platinum film can be seen in figure 4.5 as well.

In that way the platinum film was heated up to 800 K in the UHV in an oxygen atmosphere of $\approx 10^{-6}$ mbar to clean the platinum film's surface. By this cleaning procedure carbon can be removed as CO and CO_2 from the surface.

The clean platinum surface is used in the subsequent TPD studies of CO. The TPD studies were performed analogously to the studies by Campbell et al. [61]. These literature results are shown on the left hand side, whereas my results are shown on the right hand side of figure 4.6. This is the first time that a TPD study was performed characterizing the top electrode of a chemoelectronic sensor, in this case a Schottky metal–insulator (Pt/Si) system. I would like to state that we can reproduce the literature values for the CO TPD studies on Pt(111) with our chemoelectronic sensor's platinum top electrode. Nearly the same TPD spectra were recorded for the given Langmuir dose of gas (1 L = 10^{-6} torr (1.33 · 10^{-6} mbar) gas dosage for 1 s). One has to keep in mind that one has a polycrystalline platinum surface and not a Pt(111) surface.

Based on the fact that these literature values can be reproduced, one can use these data to calibrate the molecular beam with the help of more TPD studies.

4.1 Molecular beam

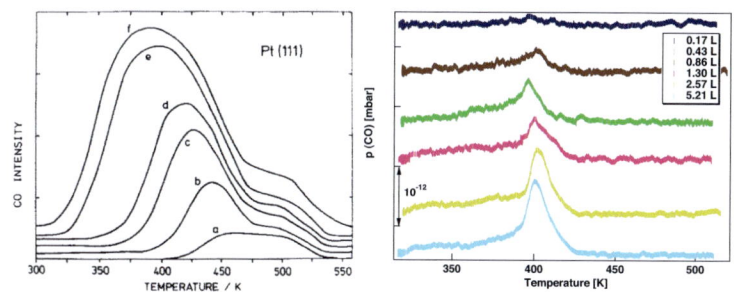

Figure 4.6: **Left hand side:** Literature TPD by Campbell et al. [61]. **Right hand side:** CO TPD on top of a platinum Schottky chemoelectronic sensor. Dosages are denoted with their value 0.17 L to 5.21 L, respectively, denoted as a to f in the TPD from literature.

For that reason the dosage of the gas is not performed with the external gas inlet system, but with the molecular beam. The molecular beam is operated with CO and a different number of pulses with 3 ms nozzle opening time (maximal opening duration) is dosed. By comparing the measured CO TPD signals (dosage with the molecular beam) with the previous signals obtained with the external gas inlet system, one can deduce from that the particle numbers. In figure 4.7 I plotted the curves for 0.43 L and 0.86 L, dosed with the external gas inlet system, together with the signal which was recorded for 45000 pulses CO with the molecular beam (1.5 bar CO on the high pressure side of the nozzle, and 2 Hz repetition frequency). Meaning with 45000 pulses we reach a coverage of around $\theta = 0.1$ (compare [61]).

This coverage value of $\theta = 0.1$ can be approximately assigned to $1.6 \cdot 10^{14}$ particles/cm². Under this simplified assumption (just comparing the desorbing species scaled with the sticking coefficient of ≈ 0.7 [59, 127, 128] one would get a particle number of the molecular beam of $5 \cdot 10^9$ particles/$(\text{cm}^2 \cdot \text{pulse})$.

To check whether really a longer nozzle opening time results in a larger observable TPD signal due to more gas molecules per pulse, the nozzle opening time was varied as plotted in figure 4.8. One can clearly see that longer nozzle opening times result in more desorbing particles, meaning more particles per pulse of the molecular beam.

However, with this method a quite small particle number of the molecular beam compared to molecular beams in the literature [67, 70, 72, 129] is found. Therefore I performed additional studies regarding a direct signal readout of the

4 Measurements, results and discussion

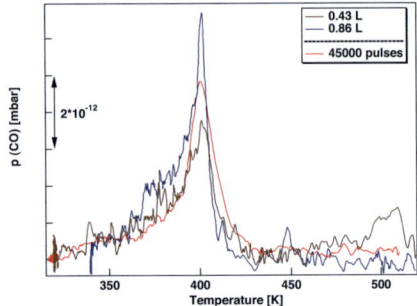

Figure 4.7: TPD curves for 0.43 L and 0.86 L dosed with the external gas inlet system and the TPD signal for 45000 pulses of CO with the molecular beam (5 bar CO on high pressure side, 2 Hz repetition frequency and 3 ms nozzle opening time).

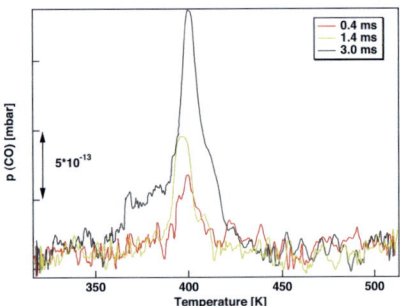

Figure 4.8: TPD curves for a dosage with 1.5 bar CO on the high pressure side of the nozzle, 2 Hz repetition frequency and 10000 pulses. The nozzle opening times are denoted in the legend.

4.1 Molecular beam

QMS to crosscheck these particle numbers of the molecular beam. In these studies the molecular beam is dosed directly into the QMS. The particle numbers of the molecular beam can then be received via an integration method of the measured QMS signal, analogously to my diploma theses [106]. With this method the particle numbers shown in table 4.1 were obtained. The rough number of pulses to obtain one monolayer of the particles (under the assumption of a sticking coefficient of 1) is given in table 4.1 as well. These numbers are comparable to previously used molecular beams [67,70,72,129]. Therefore this method of direct signal readout of the QMS seems to be better suitable to determine the particle numbers of the molecular beam.

For a wider application in chemicurrent studies it would be good to improve the molecular beam a bit further to get even higher particle numbers as 10^{15} particles/(cm$^2 \cdot$ s) [130, 131]. Then one would be able to cover the probe surface with one monolayer within one second (under the assumption of a sticking coefficient of 1). This is especially important for MIM devies, where a low chemoyield of 10^{-4} was found for the hydrogen recombination reaction on gold [7–9]. A high exposure and thus reaction rate would then still lead to measurable currents of some pA.

Gas	Particles per pulse [N/(cm^2)]	Pulses for 1 monolayer
CO	$6.9 \cdot 10^{12}$	145
H$_2$	$3.2 \cdot 10^{12}$	315
O$_2$	$7.8 \cdot 10^{12}$	130

Table 4.1: Overview over particle numbers of the molecular beam and the necessary pulses to form a monolayer (sticking coefficient = 1, sample size 1 cm^2).

4 Measurements, results and discussion

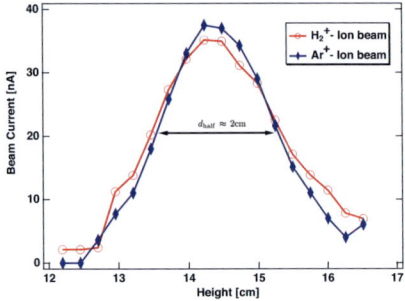

Figure 4.9: Characteristic width of a 90 eV H_2^+ and Ar^+ ion beam measured with a silicon–silicon oxide–platinum sample via the ion induced signal on platinum (measured as a current between the top platinum film and ground.

4.2 Ion beam characteristics

The ion beam is characterized regarding its width via the ion beam induced signal on platinum (measured current between the top platinum film and ground). The beam width for H_2^+ and Ar^+ ion beams (90 eV ion energy) are shown in figure 4.9. From the data one can determine an approximate beam with of $d_{\text{half}} \approx 2$ cm (taken from width at the half maximum of the ion induced signal, at the position of the sample (≈ 10 cm away from the ion source). When the ion beam is used it is important to take care to adjust for each beam energy the right beam voltage, focus voltage and deflection parameters to get a focused ion beam.

4.3 Ion migration in anodic oxide films after potentiostatic formation[1]

4.3.1 Introduction

Anodic oxidation is an intensively studied field of electrochemistry. Since the early days of so called wet diodes (based on aluminum or copper oxide [132, 133]) the technical application of thin oxide films always pushes and motivates a continuous research in this field. Today a well-founded parameter set as pH value and anion type allows potentiostatic oxidation to be a highly suitable method for the production of nonporous insulating compact oxides [88, 134–136].

Additionally, such properties as high breakdown field strength and scalability in thickness open a wide field of applications for electrochemical oxidation procedures concerning also transition metal oxides as Ti, Zr, Hf, Nb and Ta [137].

Anodic oxidation is accomplished at positive potentials. This implies an increase of the intrinsic electric field at a metal–oxide interface [138, 139] (metal–oxide interface positively charged, oxide–electrolyte interface negatively charged). The growth process is based on the conduction of ionic species supplied by the electrolyte and the electrode. Oxygen and metal ions are driven by the electric field at the metal–electrolyte interface and contribute to the observed current during the oxidation process. A simultaneous migration of metal cations and oxygen anions seems to be surprising at a first glance since the larger ion radius of O^{2-} ($r_{O^{2-}} = 132$ pm, $r_{Al^{3+}} = 57$ pm) should hinder the migration of O^{2-} anions. But defect sites in amorphous oxides provide effective transport channels also for anions [140, 141].

Wang and Hebert performed a comprehensive modern theoretical study considering the metal ion as well as the oxygen ion transport during anodic oxidation through amorphous oxide films [142]. They developed a model describing conduction as migration of defect clusters that are created by an inward displacement of oxygen ions around an oxygen vacancy. The calculations gave transport numbers for ions, which were in good agreement with experimental data indicating migration processes of both cations and anions for Al, Nb and Ta.

Oxygen vacancies also seem to play an important role in oxides formed by gas phase oxidation. Recently vacancies were found to be a reason for instabilities in aluminum oxide tunnel barriers [143].

---

[1] This section is slightly modified from its published form in J. Electrochem. Soc. 154, C663 (2007) [84].

The field driven ionic transport in the course of oxide growth can also be investigated under Ultra High Vacuum (UHV) conditions. Yates, Popova and coworkers [144–146] applied field strengths of $E = 1.4 \cdot 10^9$ V/m to a 7 Å thick oxide film. Subsequent dosing of oxygen induced a thickness increase up to 11 Å. During this additional oxide growth the initial electric field was inevitably reduced and a saturation value of the oxide thickness was reached. A maximum value of the field strength, which remains in the oxide films, is predictable. This value E_{residual} is at least smaller than $1.4 \cdot 7/11 \cdot 10^9$ $V/m \approx 0.9 \cdot 10^9$ V/m. That means that further oxide growth induced by an electric field E will certainly happen under UHV conditions with $E > E_{\text{residual}}$ when an oxygen supply is provided. For anodic oxides a similar estimation can be done. The field strength E in an oxide film has to overwhelm the formation field strength E_{form} for further anodic oxide growth. In an acetate buffer E_{form} is $0.6 \cdot 10^9$ V/m [147].

The values of E_{residual} and E_{form} are quite similar. Thus, a pronounced enhancement of ionic transport due to incorporated electrolyte species does not seem to exist in anodic oxides produced in acetate buffers (This may be different to experiments using other electrolytes [148]).

Field induced ionic transport through amorphous oxides can be understood as a hopping of ionic species between defect sites separated by a distance s. Thereby $E_{\text{residual}} \cdot s$ and $E_{\text{form}} \cdot s$ can be understood as lower limits for the activation energies of ionic transport which enables further oxide growth. Provided that s is similar in both types of oxides (gas phase and anodic), ionic transport seems to have nearly the same activation energy E_a. Assuming s as the half of the lattice constant ($s = 0.35$ nm), one would evaluate a minimum activation energy $E_a \geq 0.3$ eV.

Different values for the activation energies are found for thermally activated diffusion as self-diffusion of oxygen in aluminum and tantalum oxide. These processes are well investigated in bulk crystals. Values of $E_a = 2 - 8$ eV are found in the temperature range 1100 K − 1400 K [140, 141]. But the processes on the atomic level may be quite different in this temperature range; compared to our studies at low temperature, but high field strength.

In the present section I want to raise the question whether the existence of mobile charge carriers and the remaining field strength E_{residual} lead to diffusion processes of ions, which are influenceable by an external electric field E. In thin oxide films diffusion of cations and anions to the oxide interfaces would increase the film thickness and may change remarkably the properties of electronic sensors based on thin oxide films. Therefore a better understanding of electric field activated diffusion in thin films would help to overcome some problems in electrochemical sensor manufacturing.

An established tool for monitoring small changes in oxide films is the investigation of resistivity changes in metal–metal oxide–metal tunnel sensors [143, 149–151]. In these studies the tunnel resistance, which probes the electron transport normal to the oxide interfaces, is monitored. Since the tunnel probability is proportional to $\exp(-\sqrt{\phi_B})$ and $\exp(-d)$ (ϕ_B: height of tunnel barrier, d: oxide thickness) it is a very sensitive tool to detect even small changes in the oxide film. Additionally these samples allow the modification of the field across the oxide by applying a tunnel voltage. Konkin and Adler found a strong field dependence of the long term resistance change ($> 30\,\text{h}$) [150, 151] in such sensors. An applied positive tunnel voltage (Al ground electrode positive, Pb top electrode negative), which increases the intrinsic field in the oxide, was found to raise the tunnel resistance of the samples during a period of several 10 hours. A negative voltage that diminishes the intrinsic field, was found to keep the sensors in a stable state. The results were interpreted in terms of an ionization and subsequent migration of aluminum atoms into the oxide layer. These results would support the above-mentioned migration of mobile ions. However, the observed resistance increase can also be discussed in terms of changes in the barrier height or in the oxide thickness as well. So the results of these experiments cannot be interpreted unambiguously.

Therefore, I set up a new line of experiments to detect small deviations from the equilibrium state of the oxide at room temperature. Smooth metal films ($d = 10$ nm) were electrochemically oxidized, so that only 2 nm of the metal remained (unoxidized). A resistance measurement of the remaining metal probes the lateral electronic conductivity (see figure 4.11). Changes in the oxide may influence the resistivity of the remaining metal by two processes:

- Crystallization of the oxide would cause significant mechanical stress likely leading to a tearing of the underlying metal film. Abrupt changes of the resistivity can be expected in this case. However, at room temperatures this process is not likely since crystallization of alumina is observed at $T > 400$ K [152, 153]. For Ta_2O_5 even higher temperatures ($T \approx 600$ K) are needed [154].

- Migration of mobile ions in the oxide film into the metal or dissolution of metal atoms into the oxide would lead to a continuous oxidation of the remaining metal. A continuous increase of the metal's resistivity up to an unmeasurable infinitely high value might be detected then.

Thus, smooth resistivity changes detected as function of time are supposed to be due to ion migration.

An advantage of the new method is that the results are not influenced by the barrier height at the metal–oxide interface. Thus, the experiment is only sensitive to changes in the remaining metal. Furthermore, I combine these results with capacitance data of metal–oxide–metal systems, which probe the conduction processes normal to the oxide interface.

In the present section I apply the two mentioned experimental methods to different chemical systems:

> *i)* Ta–TaOx, since it is one of the most widely used metal–insulator systems in capacitive sensors (e.g. [155–157]).
>
> *ii)* Al–AlOx, since new theoretical studies emphasize the spin selection rules for the adsorption of oxygen on aluminum [158]. Experimental evidence for the abstraction of oxygen atoms during adsorption on aluminum was found [130]. Both the experimental and the theoretical studies seem to shed new light on this intensively investigated metal–oxide system (e.g. [159–162]).

4.3.2 Experimental

The aluminum films were evaporated on isopropanol cleaned microscope glass slides from a thermal source containing a tungsten basket mounted in a high vacuum chamber. Evaporation was done with the sample at room temperature. Tantalum films were prepared under the same conditions by an electron beam evaporator. Small deposition rates of 0.01 nm/s were chosen to ensure a homogeneous film thickness and small roughness $\approx$ 2 nm.

It should be mentioned that the film homogeneity depends on the evaporation conditions. In figure 4.10 two AFM line scans of aluminum films (15 nm nominal thickness) are shown of samples which have been prepared with deposition rates of 0.5 nm/s and 0.01 nm/s. While the latter one shows a roughness of $\approx$ 2 nm, a much higher roughness of 5 nm can be seen for higher deposition rates. The samples showing a higher roughness also have a much lower conductivity (reduced by a factor 20) compared with the smooth aluminum films, albeit the nominal thickness is the same.

Two different types of set-ups are used in the current section.

- For the resistance measurements metal films of 10 nm thickness, which were anodically oxidized up to an oxide thickness of 8 nm, were used. The 3 mm

4.3 Ion migration in anodic oxide films after potentiostatic formation

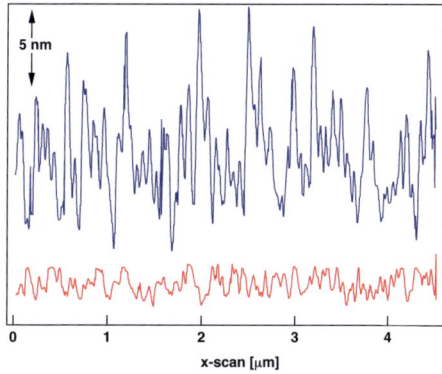

Figure 4.10: AFM line scans of differently prepared aluminum films. Upper scan taken for samples evaporated with 0.5 nm/s deposition rate. The lower one with 0.01 nm/s.

wide and 20 mm long metal samples were covered with the oxide which embraces the whole width of the metal film on a length of 5 mm, see upper half of figure 4.11

- For the capacitance measurements metal films of 30 nm thickness were anodically oxidized (on the whole width of the metal film and on a length of 8 mm) up to an oxide thickness of 2 nm. After the transfer to the UHV chamber a 15 nm thick silver film was evaporated across the oxidized area of the ground metal film, see lower half of figure 4.11

The oxidation of the metal films for both types of samples was performed in an electrolytic droplet cell as described previously in section 2.3 [85, 86] using an acetate buffer electrolyte which minimizes parallel corrosion processes during oxidation [87,88]. The aluminum films were kept under nitrogen atmosphere during the transport from the vacuum chamber to the droplet cell to avoid formation of an oxide due to ambient oxygen.

Resistivity measurements started immediately whereas capacitance measurements could be started only after 10 hours when a sufficient vacuum level ($p = 2 \cdot 10^{-8}$ mbar) was reached for the deposition of the metal top electrode. A Keithley multi-range Ohmmeter was used to measure the resistivity in DC mode.

For the capacitance measurements the silver top electrode was taken as a ground and the bias voltage was applied to the aluminum or tantalum back electrode. Hence, a positive bias voltage means an increase of the remaining field strength

4 Measurements, results and discussion

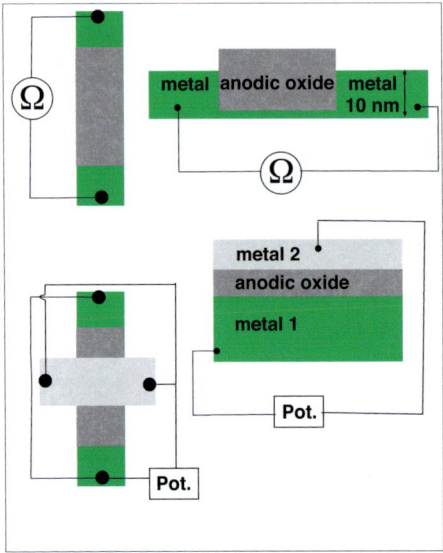

Figure 4.11: **Upper half:** Setup for resistance measurements. The aluminum and tantalum stripes were oxidized up to a remaining metal thickness of 2 nm. **Lower half:** Setup for capacitance measurements. The aluminum respectively tantalum films are oxidized up to a thickness of 3.1 nm. The top metal electrode is evaporated in cross geometry onto the oxidized base electrode. Pot. stands for the connected potentiostat.

$E_{residual}$ whereas a negative bias voltage weakens it. Capacitance measurements were always performed at 0 V bias voltage by monitoring the charging current when a triangular voltage of 0.1 V with a scan rate of $dU/dt = 0.02\,V/s$ was applied. Experiments dealing with the temporal changes of the capacitance under applied bias voltage always show values taken at 0 V. The bias voltage was interrupted for the measurement period (10 s).

The prepared oxide thicknesses were confirmed ex situ by TOF-SIMS (Time of Flight Secondary Ion Mass Spectroscopy) and XPS-Sputter-profile measurements.

4.3.3 Results

4.3.3.1 Electrochemical oxidation

A typical oxidation cyclovoltammogram (further called CV) of a freshly evaporated aluminum–film is shown as the upper graph of fig. 4.12. Starting with a cathodic current density $j = -20\,\mu Acm^{-2}$ at $E_{SCE} = -1.1\,V$ the current increases up to a plateau current density $I_{pl.}$ of 300 μAcm^{-2}. In the cathodic sweep the current rapidly drops to zero.

The CV changes when the sample is exposed to a dry oxygen atmosphere for two hours instead of a nitrogen atmosphere ($p(O_2) = 10^5$ Pa), see middle graph. The cathodic current density at $E_{SCE} = -1.1\,V$ vanishes, the current remains at low values ($< 10\,\mu Acm^{-2}$) up to a potential of $E_{SCE} = -0.4\,V$. Then the current increases and shows a pronounced overshoot. The overshoot is followed by a plateau current, which has the same value as in the previous experiment. The current overshoot is usually explained by the sequence of two processes:

> *i)* At first emission of O^{2-} and Al^{3+} ions starts from the oxide–electrolyte and metal–oxide interfaces, respectively.
>
> *ii)* Subsequently the charge carriers meet the opposite interface, further oxide growth starts and the current settles down to I_{pl} again [163–165].

The delayed onset of the plateau current density ($\Delta E \approx 0.8\,V$) of the gas phase oxidized sample compared to the freshly prepared sample can thus be taken as a measure for the thickness of the gas phase oxide layer d_{gpo}. It would correspond to $d_{gpo} = 0.8\,V \cdot E_{form} = 1.4\,nm$ ($E_{form} = 1.6\,V/m$ [147, 166]). This value differs from that one found previously for polished wires ($d_{gpo} = 2.5$ nm) [166]. We think that this difference might be due to the higher defect density of polished wires. A value of 2.5 nm could also be reproduced for metal films, which were evaporated

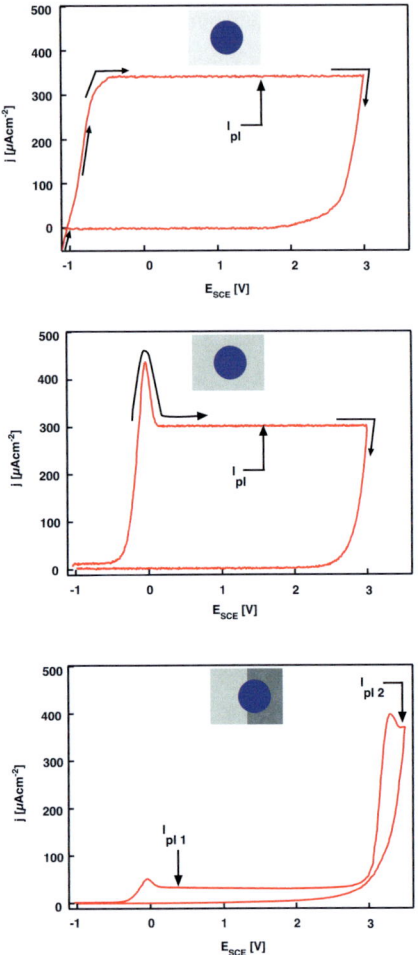

Figure 4.12: **Upper graph:** Cyclovoltammogram of a freshly prepared aluminum film ($E_{initial} = -1.1\,V$, $E_{reverse} = +3.0\,V$, $dE/dt = 0.1\,V/s$). Droplet cell positioned in the center of the film as shown in the inset. **Middle graph:** Cyclovoltammogram of an aluminum film exposed for two hours to a dry oxygen atmosphere ($p(O_2) = 10^5$ Pa). **Bottom graph:** Droplet positioned on the intersection of anodically oxidized ($E_{final} = 3.0\,V$) and ambient gas phase oxidized aluminum film (see inset). Cyclovoltammogram recorded up to $E_{final} = 3.5\,V$.

with a higher deposition rate of 0.1 nm/s. These samples had a higher roughness (5 nm) than the ones evaporated with 0.01 nm/s ($\approx$ 2 nm) as determined by AFM measurements. The results show that the final thickness of gas phase oxide layers decreases with decreasing defect density of the aluminum surface. For single crystals even smaller values of 0.4 nm were reported [167].

The potential of the current overshoot can also be taken as a tool to study inhomogeneous oxide structures. An inhomogeneous structure was produced in the following manner: half of the area of the metal film was held under nitrogen atmosphere. Then the other half was oxidized under the droplet cell by a CV with $E_{final} = +3.0$ V. Then the whole sample was exposed to dry oxygen atmosphere for 2 hours. Afterwards the droplet cell was positioned at the intersection of the gas phase oxidized part of the aluminum film and the anodic oxide. The subsequent CV with $E_{final} = +3.5$ V shows two discriminable current overshoots corresponding to the previously established oxide thicknesses of the gas phase and the anodic oxide. The ratio of the plateau current values I_{pl_1}/I_{pl_2} mirrors the area ratio $a_{gpo}/a_{gpo}+a_{anod}$ of the two oxide types covered by the droplet cell.

Oxidation of tantalum delivers comparable CV. The overshooting current during further anodic oxide growth can also be used in this case to determine the previously established oxide thickness.

4.3.3.2 Capacitance measurements

In figure 4.13 capacitances measurements of aluminum–aluminum oxide–silver samples are presented for different bias voltages. The traces show a pronounced dependence on the bias voltage. A negative bias voltage of $U_T = -1$ V leaves the capacitance of the sample in a long term stable state. Zero volt bias voltage leads to a steady loss of capacitance over the whole time. A stable value does not appear even for $t > 1 \cdot 10^5$ s. A positive bias of $U_T = +1$ V on the other hand leads to a quite fast steady state value for $t > 6 \cdot 10^4$ s.

The observed capacitance losses with positive and zero bias voltages are irreversible and cannot be nullified by a negative U_T. Considering the dipole layers at the two oxide interfaces one can easily imagine field induced exchange processes between the two interfaces. However, this process should be reversible. The non-reversibility points more to an either permanent decrease of the dielectric permittivity ϵ_r or to a slight thickness change of the oxide. Taking ϵ_r as unchanged during the experiment, one can evaluate a maximum value of the thickness change $l(t)$ simply by:

4 Measurements, results and discussion

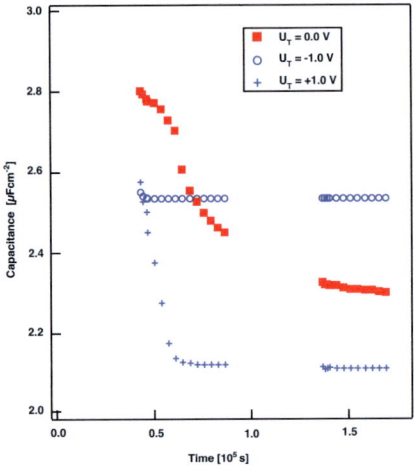

Figure 4.13: Capacitances measured at ($p \approx 1 \cdot 10^{-9}\,\text{mbar}, T = 300\,\text{K}$) of aluminum–aluminum oxide–silver samples (oxidation potential 1.0 V, initial oxide thickness 3.1 nm) plotted as function of time after the end of the anodic oxidation. At t = 43,300 s the deposition of the silver top electrode was completed. Traces of three different samples studied at different bias voltages are shown.

4.3 Ion migration in anodic oxide films after potentiostatic formation

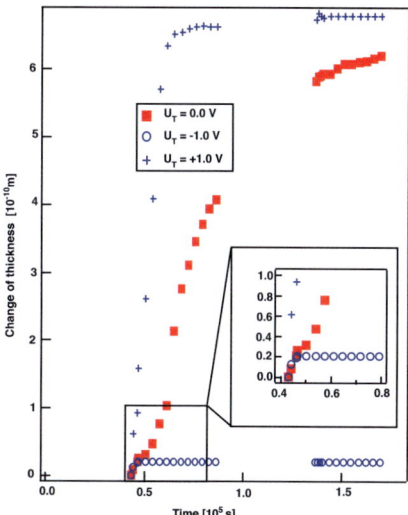

Figure 4.14: Calculated thickness changes of the aluminum oxide layer derived by equation 4.1 and the experiments in fig. 4.13. Inset shows a zoom on the change of thickness $l(t)$ for $4 \cdot 10^4\,s < t < 8 \cdot 10^4\,s$.

$$C(t) = \epsilon_0 \cdot \epsilon_r \frac{A}{d_0 + l(t)}. \qquad (4.1)$$

$l(t)$ is evaluated from the measurements shown in figure 4.13 and plotted in figure 4.14.

A maximum thickness change of 6 Å is calculated which corresponds to around 2 monolayers (lattice constant of the oxygen fcc-lattice in γ alumina is 2.8 Å [168]).

However, a clear assignment of the capacitance changes to a thickness increase cannot be suggested on the basis of the data presented here. The data could also be explained by a reduction of ϵ_r by 40 %.

Corresponding experiments with tantalum–tantalum oxide sensors show only constant capacitance values. The results also do not depend on the bias voltage up to $\pm 1\,V$ in this case and are therefore not shown here.

4 Measurements, results and discussion

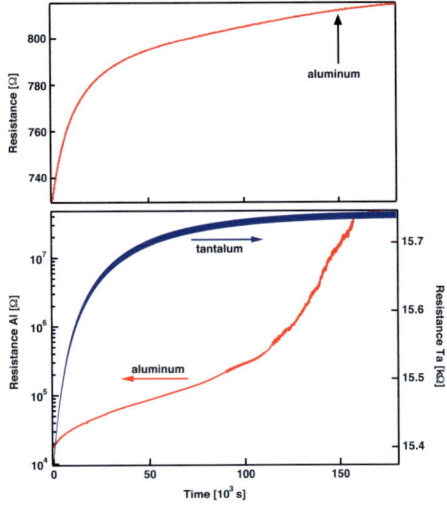

Figure 4.15: **Upper graph**: Resistance change of a 10 nm aluminum film (deposited with a rate of 0.01 nm/s) of which the top 6 nm have been oxidized (a 4 nm thick aluminum film remains). Similar experiments with tantalum films show resistance changes below 0.1 % and are not shown here. **Lower graph:** Same experiment for 2 nm thick remaining metal films after preparation of 8 nm thick oxides as top layer. Values for the tantalum film are added as second y-axis. Experiments performed at $T = 300$ K.

4.3.3.3 Resistance measurements

In figure 4.15 the resistances R of 4 and 2 nm thick metal films (aluminum and tantalum respectively) are plotted versus time.

For 4 nm thick tantalum R remains constant at $R = 2500 \, \Omega$ from $t = 0$ s to $t = 1.8 \cdot 10^5$ s whereas R for 4 nm thick aluminum increases noticeably from 730 Ω to 815 Ω on the same time scale, see upper graph of figure 4.15. A saturation value of 820 Ω is reached for $t > 3 \cdot 10^5$ s (not shown in fig. 4.15).

The different behavior of aluminum and tantalum is even more pronounced for 2 nm thick metal films. The resistance of the tantalum film increases slightly from 15.4 kΩ to 15.8 kΩ, see lower graph. The aluminum film exhibits a permanent increase of the resistance. The rate dR/dt increases again at $t = 1.2 \cdot 10^5$ s. At $t = 1.8 \cdot 10^5$ s R reaches a value of $4 \cdot 10^7 \, \Omega$ which is the detection limit of our measurement system. Similar observations could be reproduced for several aluminum samples but a comparable resistance increase for tantalum samples could never be found.

It should be mentioned that 2 nm thick tantalum films loose their conductivity completely when the thickness of the oxide top layer is increased by 0.4 nm. Hence, tantalum films with thicknesses below 1.5 nm seem to be insulators. Thus I can exclude thickness changes in the range of several 0.1 nm for the explanation of the resistance changes for tantalum shown in the lower graph of fig. 4.15.

One can conclude that the resistance experiments point to a thickness increase of the anodic oxide in the case of aluminum after the end of potentiostatic oxidation. One could thus assign the observed temporal behavior of the capacitances in aluminum oxide capacitors to thickness changes. This will be further discussed in the next section.

Capacitance and resistance measurements concordantly point to a stable metal––oxide interface for tantalum–tantalum oxide systems.

4.3.3.4 The influence of film homogeneity on the film resistance

The already mentioned influence of the deposition rate on the structure of the evaporated aluminum films on the nano scale (see figure 4.10) has also an important impact on the oxidation reaction of the films. In figure 4.16 the development of the resistance of a rough aluminum film (see figure 4.10) is shown after anodic oxidation has been terminated. The oxidation potential was set to 3.6 V. The remaining metal thickness is estimated to be approximately 4 nm thus giving approximately the same initial resistance value as in the experiment with the smooth aluminum film discussed above . The initial phase of resistance increase also seems to similar in the two experiments. But after 10^4 s a sharp increase by a factor of 1.9 occurs within a period lasting only several seconds. Then the resistance increase continues in the same manner as before. The phenomenon of a sharp resistance increase appears in this experiment again two times, leading to a total conductivity loss after $8.1 \cdot 10^4$ s. In a series of experiments with rough aluminum films similar events of sharp resistance increases could be observed. But neither the time nor the amount of resistance increase could be reproduced. This points to the statistic nature of these processes on rough aluminum films.

These sudden events of sharp resistance increases are a special property of rough aluminum films. They may be explained by processes as ripping and off peeling. However, the steady continuous resistance increase documented in figure 4.15 cannot be explained by these processes.

4 Measurements, results and discussion

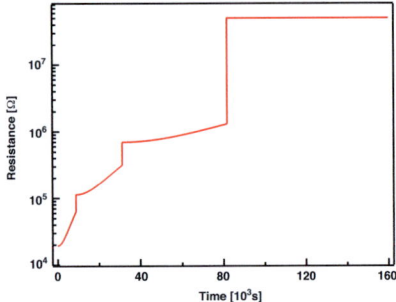

Figure 4.16: Resistance change of a 10 nm aluminum film (deposited with a rate of 0.5 nm/s) of which the top 6 nm have been oxidized.

4.3.4 Discussion

The experiments shown above point to small thickness changes after potentiostatic oxide formation of the aluminum oxide (see figure 4.14). It could be explained by either a migration of aluminum atoms into the oxide or by oxygen ions penetrating the aluminum film underneath. Both processes would be enhanced by an applied positive bias voltage in agreement with our experimental results. Keeping in mind the big difference in melting points for Al and Ta [169] ($T_{melt}(Al) = 933\,K$, $T_{melt}(Ta) = 3290\,K$) one can easily imagine, that migration of Ta atoms into the oxide proceeds at a much lower rate than the same process for Al.

Migration of oxygen ions to the oxide surface might also occur. It can be detected by the oxidation of adjacent metal layers (see upper graph of figure 4.17). The oxidation of evaporated copper films (up to 1 monolayer copper oxide) was reported for polycrystalline aluminum oxide [170] as well as on $Al_2O_3(0001)$ single crystal surfaces [171] both prepared in UHV. Albeit the migration of oxygen ions into silver is not supported by a positive bias, one could speculate that the small capacitance loss, which appears even at a negative bias voltage (see zoom in figure 4.14), might be due to that process.

This observation might also be due to a more or less instantaneous chemical reaction of the metal with the surface layer of the oxide. On anodic oxide surfaces hydroxide layers can be found [172, 173]. This may provide reactants for an oxidation reaction of the top metal M_{top} by $2\,OH^- + M_{top} \rightarrow H_2 + 2\,O^{2-} + M_{top}^{2+}$. This reaction was found to be important for Cr, Fe, Co, Ni, but less significant for

4.3 Ion migration in anodic oxide films after potentiostatic formation

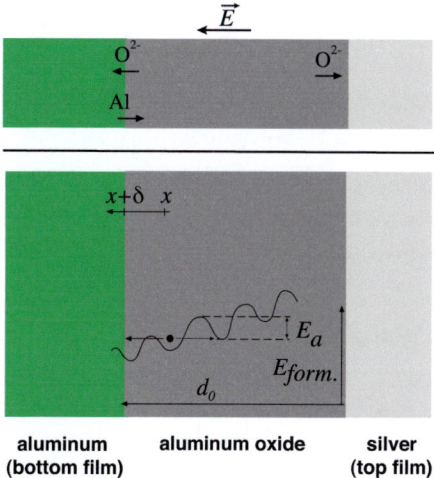

| aluminum | aluminum oxide | silver |
| (bottom film) | | (top film) |

Figure 4.17: **Upper graph:** Schematic drawing of possible migration processes. **Lower graph:** Activated (E_a: activation energy) hopping processes of ions in the oxide along the electric field $E_{\text{form.}}$.

the noble metals [174].

Since our resistance measurements show clearly that the main processes occur at the bottom interface, our results will be discussed in the frame of the two following processes (both occurring at the aluminum–aluminum oxide interface):

1. Al-atom migration into the oxide

2. excess oxygen migration from the oxide into the Al layer.

At first a simple macroscopic description will be analyzed. Then two more detailed models considering the above mentioned processes will be discussed.

4.3.4.1 Macroscopic description

The initial oxide thickness at $t = 0\,\text{s}$ after oxide formation is d_0. During the thickness increase ($\dot{d}(t) \neq 0$) the field strength E_{oxide} in the layer (initial value E_{residual}) decreases $\propto 1/d$. From a simple point of view one can assume a proportionality between the instantaneous field strength $E_{\text{residual}} \cdot d_0/d(t)$ and the growth rate $\dot{d}(t)$, thus giving:

4 Measurements, results and discussion

$$\dot{d}(t) = A \cdot E_{\text{residual}} \cdot \frac{d_0}{d(t)} \quad (4.2)$$

A is a proportionality constant ($[A] = [m^2/\text{V·s}]$). To obtain a value of A, I interpret the resistance measurements in figure 4.15 in the following way:

The oxide thickness after formation was 8 nm ($d_0 = d(t = 0\,\text{s}) = 8\,\text{nm}$). E at $t = 0$ s is taken to be E_{residual}. The remaining aluminum lost its resistance completely at $t_{\text{end}} \approx 1.5 \cdot 10^5$ s. Thus I set $d(t_{\text{end}}) = 10\,\text{nm}$.

The solution of equation 4.2 is

$$d(t) = \sqrt{2 \cdot A \cdot E_{\text{residual}} \cdot d_0 \cdot t + d_0^2}, \quad (4.3)$$

leading to $A = 2.5 \cdot 10^5\ m^2/\text{V·s}$. Equation 4.3 delivers the well known square root law. But saturation values as found in the experiment do not match with equation 4.3. Hence, one has to look for different models.

4.3.4.2 Microscopic description: Migration of oxygen ions

The hopping of ions in amorphous oxides can be treated as transport between two local potential minima (see lower graph of figure 4.17). We consider a system where a non-stoichiometric ion can hop between two lattice sites with an activation energy E_a. Due to the remaining field strength in the oxide one has to consider a hop along an asymmetric potential scheme [175]. Thereby one has to balance an uphill transition rate against the field strength E_{oxide} and a downhill transition rate with the field strength. The uphill transition rate against the field strength E_{oxide} may then be written as [176]:

$$k_\uparrow = \nu \cdot e^{\frac{-(E_a + e \cdot \delta \cdot E_{\text{oxide}})}{k_B \cdot T}}$$

E_a is the activation energy for the hop along a distance δ, ν can be understood as the attempt frequency for the hopping event. In our experiments the field strength after the potentiostatic oxidation will decrease due to the slight thickness increase. Thus, we have to consider a time dependent transition rate for the uphill transition rate $k_\uparrow(t)$, which can be formulated as follows:

$$k_\uparrow(t) = \nu \cdot exp\left(\frac{-E_a + e \cdot \delta \cdot E_{\text{oxide}} \cdot d_0/d(t)}{k_B \cdot T}\right) \quad (4.4)$$

4.3 Ion migration in anodic oxide films after potentiostatic formation

Downhill the transition rate is:

$$k_\downarrow(t) = \nu \cdot exp\left(\frac{-E_a - e \cdot \delta \cdot E_{oxide} \cdot d_0/d(t)}{k_B \cdot T}\right) \quad (4.5)$$

The downhill density n_d will populate and the uphill density n_u depopulate according to first order rate equations:

$$\begin{aligned}\dot{n}_u &= k_\uparrow(t)n_d - k_\downarrow(t)n_u \\ \dot{n}_d &= k_\downarrow(t)n_u - k_\uparrow(t)n_d \end{aligned} \quad (4.6)$$

The initial conditions for the ion densities at $t = 0$ s directly after oxide formation are named n_u^0 and n_d^0. For n the solution of the coupled differential equations including the difference of forward and backward reactions one can derive $n_d(t) - n_u(t)$ as:

$$\begin{aligned}n_d(t) - n_u(t) = \frac{1}{k_\downarrow(t) + k_\uparrow(t)}\Big[&-k_\uparrow(t) \cdot (n_d^0 - n_u^0) \\ &-2 \cdot e^{-(k_\downarrow(t) + k_\uparrow(t)) \cdot t}(k_\downarrow(t) \cdot n_u^0 - k_\uparrow(t) \cdot n_d^0) \\ &+ k_\downarrow(t) \cdot (n_d^0 + n_u^0)\Big] \end{aligned} \quad (4.7)$$

Multiplying the above term with the oxide thickness d_0 gives the number of ions N_{ion} which leave the oxide per unit area. As a simple approximation I take $3/5 \cdot N_{ion}$ to be oxygen ions which migrate into the aluminum film. Three oxygen ions may reduce two aluminum ions. The number of Al atoms per m^2 in one (111) layer, N_{Al}, can be estimated as

$$N_{Al} = a_{Al} \cdot \sqrt{3} \cdot \frac{\rho(Al)}{M_{mol}(Al)} \quad (4.8)$$

with $\rho(Al)$: density, $M_{mol}(Al)$: molar weight, $a(Al)$: lattice constant of the Al.

The number of oxidized aluminum layers, l, is then $\frac{2}{5} \cdot N_{ion}/N_{Al}$. $l(t)$ can finally be written as:

$$l(t) = \frac{2}{5} \cdot d_0 \cdot \frac{n_d(t) - n_u(t)}{a(Al) \cdot \sqrt{3} \cdot \frac{\rho(Al)}{M_{mol}(Al)}} \quad (4.9)$$

4 Measurements, results and discussion

where d_0 denotes the thickness of the oxide layer at $t = 0$ s. Since $k_\downarrow(t)$ and $k_\uparrow(t)$ are functions of time and thereby depend on the oxide thickness, the equation system 4.4 - 4.9 cannot be solved analytically. The equations are thereby iteratively solved on a time mesh with $t = 0.1$ s resolution. Densities of mobile ions $n_d^0 = n_u^0 = 0.5 \cdot 3/5 \cdot \rho_{Al_2O_3}$ were used according to Möhring, who determined the amount of mobile species by modelling the current overshoot in the CV of anodic oxidation [165]. In equations 4.4 and 4.5 the hopping distance δ is taken to be a half of the oxides lattice constant a ($a = 2.8 \cdot 10^{-10}$ m) [168] ($\delta = 1.4 \cdot 10^{-10}$ m). The attempt frequency ν can be taken from the phonon frequency of the oxide. Characteristic phonon frequencies are measured at 400 cm^{-1} and 960 cm^{-1} by EELS (electron energy loss spectroscopy) in non-stoichiometric oxide films [177–179] corresponding to around 10^{12} s^{-1} for ν. Plots of the change of the thickness, $l(t)$, of the aluminum exemplify three different values of the activation energy E_a (see lower graph of figure 4.18). A saturation value (further called l_∞) of $l_\infty \approx 4$ Å is calculated. The activation energy determines the time dependence until the saturation value is reached. The experimental limitations (start and end of data recording $t > 10$ s, $t < 3 \cdot 10^5$ s) restrict the comparison with the theoretical model to activation energies for which 0.9 eV $< E_a < 1.2$ eV. The corresponding depletion of mobile charge carriers in the oxide is shown in the upper graph of figure 4.18.

The derived values for E_a from 1.0 to 1.2 eV are $\approx 50\%$ smaller than the values of the activation energy for oxygen self-diffusion (≈ 2.2 eV) in polycrystalline alumina [180].

Albeit the maximum value of the thickness increase at $t = \infty$, l_∞ is close to the experimentally found value of 6 Å, one has to discuss several shortcomings of the model:

i) The maximum thickness increase l_∞ scales linearly with the initial oxide thickness d_0 (see equation 4.9). This is not confirmed by capacitance measurements. The evaluation of the capacitance changes ΔC for samples with different initial oxide thicknesses d_0 is listed in table 1. An evaluation with equation 4.1 leads to more or less constant thickness changes l_∞. This is not consistent with equation 4.9.

ii) The evaluated maximum values of l_∞ in fig. 4.14 can only be explained by reasonable values of δ, E_{oxide} and E_a when an unrealistically large amount of mobile species of $n_d^0 = n_u^0 = 0.5 \cdot 3/5 \cdot \rho_{Al_2O_3}$ is assumed.

4.3 Ion migration in anodic oxide films after potentiostatic formation

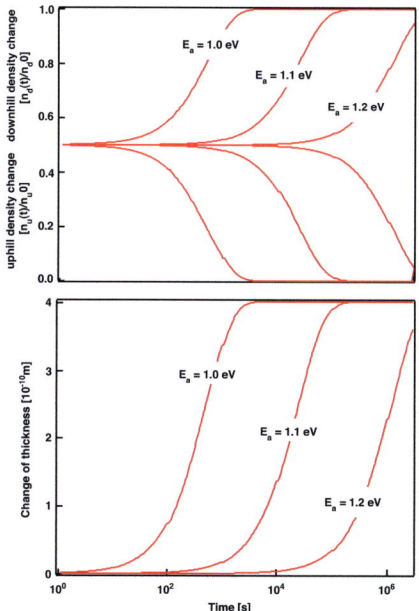

Figure 4.18: **Upper graph:** Calculated depletion of mobile charge carriers $n_u(t)$ and $n_d(t)$ for the same parameters as used in equations 4.4 - 4.9. **Lower graph:** Calculated thickness change, $l(t)$, of the aluminum plotted vs. logarithm of time according to equations 4.4 - 4.9, for different values of the E_a; $E_{\text{oxide}} = 9 \cdot 10^8 \frac{V}{m}$, $n_d^0 = n_u^0 = 0.5 \cdot \rho_{Al_2O_3}$, $T = 300$ K and $d_0 = 8$ nm.

Therefore a different approach has to be considered.

d_0 [nm]	ΔC [μFcm^{-2}]	l_∞ [Å]
2	0.7	5
3	0.31	4
3.9	0.19	3
4.8	0.13	4

Table 4.2: Capacitance changes ΔC of samples with different initial oxide thickness d_0 measured at $U_T = 0$ V. l_∞ is evaluated by equation 4.1 with a constant ϵ_r.

Aluminum-vacancies are likely to be found on γ - alumina interfaces [181]. In our case an aluminum-deficient amorphous oxide could induce an ionization of aluminum atoms and a subsequent migration of cations into the oxide film. This would lead to a consumption of the aluminum layer lying below the oxide and a slight oxide growth.

4.3.4.3 Microscopic description: Aluminum cation migration

An ionization of aluminum atoms and a subsequent migration of aluminum cations could depend on the electric field strength applied to the sample. The electric field resident in the oxide film will influence the first layers of the aluminum film since the screening length in the metal film is of the order of the lattice constant [182]. A field dependent "melting" or "viscous flow" [183] of one monolayer metal atoms and subsequent solution into the oxide is thus conceivable.

The formulation of field dependent transition rates is equivalent to that one presented in equations 4.4, 4.5, 4.7. As $n_u(t = 0\,\text{s})$ we take the areal density of aluminum atoms in one monolayer N_{Al} as $n_u^0 = 4.2 \cdot 10^{19}$ atoms $\cdot$ m^{-2}. The state n_d^0 is taken to be the Al^{3+} ion state in the aluminum–oxide interface which is assumed to be empty at $t = 0\,\text{s}$. The jump distance (δ in equations 4.4 and 4.5) from the metal substrate into empty interstitial sites in the interface is taken to be $\delta = 1.2$ Å [167].

In figure 4.19 the calculated depletion of aluminum is plotted for a set of tunnel voltages U_T from -3 V to $+1$ V for two different activation energies $E_a =$

4.3 Ion migration in anodic oxide films after potentiostatic formation

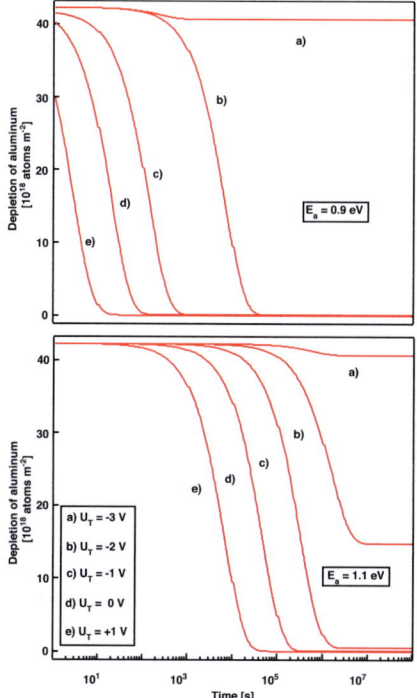

Figure 4.19: **Upper graph:** Calculated depletion of aluminum in contact with amorphous oxide (jump distance $\Delta = 1.4\,\text{Å}$) plotted for different tunnel voltages across a 2.4 nm thick oxide layer. Activation energy $E_a = 0.9\,\text{eV}$. **Lower graph:** Same calculation as above with $E_a = 1.1\,\text{eV}$. Calculations performed for $T = 300\,\text{K}$.

0.9 v. 1.1 eV (upper and lower graph). The activation energy determines the temporal behavior of $n_u(t)$. This is similar to the calculation presented in figure 4.18.

The experimentally observed bias voltage dependence of the capacitance change is partly reflected by the calculation. A negative bias voltage of $-3\,\text{V}$ is needed to diminish the depletion of n_u whereas only $-1\,\text{V}$ were needed to keep the capacitance stable in the experiment. This discrepancy might be caused by an inhomogeneous potential distribution. Especially at the interfaces the potential drop might be more pronounced than in the bulk of the oxide.

4.3.5 Conclusion

Recapitulating I can state that field strength dependent migration processes modify the properties of anodic oxides even in Ultra High Vacuum after the end of potentiostatic oxidation. However, these effects are much more significant for aluminum than for tantalum. This big difference is surprising since the electrochemical properties (cyclovoltammogram, migration of both ion types during growth, film formation factor [147]) are quite similar. After the end of potentiostatic oxidation both oxides will contain mobile cation and anion species [142]. These results stress the necessity to discuss ageing processes in terms of both anion and cation migration.

A simple macroscopic description gives a square root law, which does not explain the observed saturation values for the thickness increase.

A microscopic description concerning the migration of anions (oxygen interstitials) and subsequent oxidation of the bottom metal contains several deficiencies (high amount of several 10 % for concentration of mobile species, proportionality to initial oxide thickness). But, activation energies E_a in the range from 0.9 eV to 1.2 eV were found to be consistent with the temporal evolution of the capacitance as well as the resistance measurements. Additionally the calculated proportionality of the oxide grown after the end of potentiostatic oxidation to the initial oxide thickness is not in concordance with the experiments.

A microscopic description considering the migration of metal cations into the oxide is in accordance with the experimental data for the same activation energies as in the anion migration model. The cation migration can be justified by metal vacancies at the metal–oxide interface (in the case of aluminum oxide) and is thus independent from the oxide thickness.

The simplest argument rationalizing the difference between tantalum and aluminum found in the experiments is the difference in the melting points of the metals. For aluminum atoms one could speculate about a lower activation en-

4.3 Ion migration in anodic oxide films after potentiostatic formation

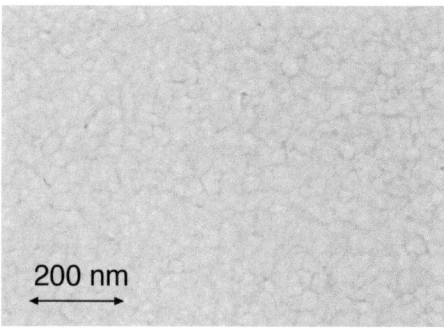

Figure 4.20: SEM image of an oxidized aluminum film on a silicon substrate.

ergy for a hopping process from the metal into the oxide compared to tantalum.

Moreover one difference might be the absence of tantalum vacancies at the tantalum–oxide interface. This would hinder the diffusion of further metal atoms into the oxide.

4.3.6 Appendix

In figure 4.20 a SEM image of an oxidized aluminum film on a silicon substrate can be seen. No rips or breaks can be observed.

4.4 Preparation and properties of thin amorphous tantalum films[2]

4.4.1 Introduction

Thin tantalum films and contacts may play a key role in future micro- and nano-electronic sensors due to their high stiffness and corrosion resisitance [184–187]. For that reason, the preparation of thin tantalum films seems to be of continuing interest despite the large amount of literature already published in the last four decades [188–195]. These earlier studies were mainly focussed on the macroscopic properties of the films. A comprehensive review of the early period, handling in particular the main preparation methods sputtering and electron evaporation with bent beams, can be found in Westwood's book [196].

The interesting and sometimes confusing point concerning tantalum as well as niobium and tungsten films is the allotropy. These films appear in simple body centered configurations as well as in tetragonal phases [197]. Also phase transitions may appear [198]. Both configurations of tantalum can be produced in sputter deposition processes. A nearly completely amorphous modification of tantalum films can be produced by pulsed laser depositon [199].

The different phases of tantalum show distinct differences in their electrical resistivity:

- Low resistivity (10 - 60 $\mu\Omega \cdot$ cm) α−tantalum in the body centered cubic phase [200, 201]

- High resistivity (100 - 200 $\mu\Omega \cdot$ cm) β−tantalum in the tetragonal phase [195, 197, 202, 203]

- Completely amorphous films ($> 200\,\mu\Omega \cdot$ cm) [199].

The crystalline modifications of thin tantalum films can be used for bonding and electric interconnects whereas the amorphous modification takes on a special position. The amorphous modification's application field is more located in mechanical applications due to their peculiar elastic strain limit and fracture toughness [204].

I would like to present in this section evidence that an amorphous modification of tantalum can be expected, when films are e-beam evaporated on simple glass substrates. The appearance of the amorphous structure is surprising since

---

[2]The majority of this section was published as an article in J. Phys. D 42, 135417 (2009) [82].

comparable experiments with sputtered tantalum films on glass substrates lead to polycrystalline samples [184, 205]. The difference may be caused by the existence of ions and particles with higher kinetic energy in the sputter process [206], whereas in the e-beam set-up presented here, a large sample-source distance of around 20 cm suppresses the contribution of ions due to their strong broadening (ions more than $\frac{1}{r^2}$, atoms $\approx \frac{1}{r^2}$).

The structure of the samples is investigated by Scanning Electron Microscope (SEM) and Atomic Force Microscope (AFM) investigation. These techniques reveal that our films show a continuous smooth structure and suggest an amorphous modification of the tantalum.

However, the discrimination between a nano crystalline and an amorphous structure is difficult by means of microscopy techniques only. Therefore resistivity measurements as function of film thickness are used to characterize the overall conduction mechanism up to thicknesses of 100 nm. These experiments are combined with measurements of the temperature coefficient of the resistivity (TCR). This method was often applied to nitride and oxide films of tantalum [184, 207]. In the experiments presented in this section the method is applied to an elemental tantalum film to determine the amorphous structure of the tantalum film in the context of Tsuei's and Mooij theory of electronic transport in strongly disordered metals [208, 209].

4.4.2 Experimental

4.4.2.1 Electron beam evaporation

A commercial e-beam evaporator (TECTRA) is used in our experiments. The filament (tungsten wire of 0.5 mm thickness), is heated by a current density of $2 \cdot 10^3$ A/m^2. Between the filament wire and the tantalum rod (distance ≈ 5 mm) a voltage $U = 1.3$ kV is applied. The resulting field strength E can be estimated to $E = 1.3$ kV/5 mm $\approx 2.6 \cdot 10^5$ V/m which is clearly below the breakdown field strength $E_{DB} = 1.5 - 2.6 \cdot 10^8$ V/m of tungsten filaments under vacuum conditions [210]. The electron current emitted from the filament is measured at the tantalum rod (emission current).

The 3 mm thick rod forms a tantalum sphere with a radius of 3 mm at its top (see fig. 4.21). Under the simplifying assumption of a homogeneous electron current on the hemisphere of the molten tantalum one can estimate the necessary current density j_e for the evaporation by the ratio of the emission current I_e during evaporation and the hemisphere's surface (see fig. 4.21) by $j_e = \frac{I_e}{\frac{1}{2} \cdot 4 \cdot \pi \cdot r^2}$ to $2 \cdot$

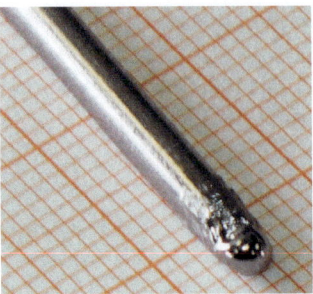

Figure 4.21: Photo of the tantalum rod. The hemisphere (diameter ≈ 3 mm) due to the evaporation can be seen on the right side at the top of the rod.

10^3 A/m².

Since the voltage on the tantalum is kept constant, the dissipated power is simply $P = U \cdot j_e$, which has a value of $2.6 \cdot 10^6$ W/m² in our case. The vapor pressure p in Pa of the tantalum [196, 211] is given by the temperature T in Kelvin:

$$p = 10^{11.1-(47000 \cdot K/T)} \text{ Pa} \qquad (4.10)$$

The path of the evaporating tantalum atoms intersects with the path of the electrons. This leads to a partial ionization of the atomic beam. Some of these ionized atoms impinge the flux electrode, which is an isolated electrode mounted on the front end of the e-beam evaporator. The tantalum ions cause a small current, which is in the range of $1 - 2$ µA. Assuming a constant ionization rate, a control of the tantalum atom flux is possible by monitoring the tantalum ion current at the flux electrode. The distance between the top of the flux monitor electrode at the end of the electron beam evaporator and the substrate surface is 178 mm.

The diameter of the flux electrodes aperture is 6 mm, which is 3 mm larger than the diameter of the tantalum hemisphere. There is a clear line-of-sight from the whole hemisphere to the substrate. Thus, a simple treatment of the evaporator as a Knudsen type effusion cell is not sufficient. A derivation of the thickness distribution caused by ample evaporation sources (planar and hemispherical sources) will be given in the appendix.

A quartz micro balance is mounted just besides the sample. Figure 4.22 shows the sample holder set-up with the quartz micro balance mounted nearby. One can nicely see the corona like deposited gold and tantalum which was evaporated in subsequent experiments. The substrates were isopropanol cleaned microscope

4.4 Preparation and properties of thin amorphous tantalum films

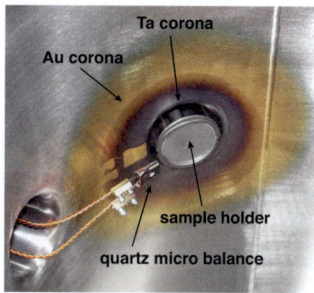

Figure 4.22: Photo of the sample holder, quartz micro balance and evaporation corona of gold and tantalum.

glass slides. The deposition rate was fixed to 0.4 nm/minute for all samples.

4.4.2.2 Tantalum film characterization

Characterization of the tantalum film by SEM

In fig. 4.23 SEM images of the evaporated films with 20 nm and 100 nm thickness are shown. No structures are observable on the 20 nm sample. Small structures (approximate size 5 − 10 nm) can only be seen on the 100 nm sample. We did not investigate thicker films of for example 500 nm thickness. It would be interesting whether there is a thickness induced change towards a more polycrystalline structure. A thickness induced change of sample structure cannot be excluded since Zhou and coworkers observed larger structures and grains on 500 nm thick tantalum films [184]. But these samples contained 20 atom % oxygen. The oxygen contents of our samples were determined by TOF SIMS (Time of flight secondary ion mass spectrometry) to be below 1 % [212].

Hence, I think that especially for the thickness range ≤ 40 nm our images point to a homogeneous (maybe even amorphous) structure with a small roughness which contains elemental tantalum only. The homogeneous structure will be verified in the next section by AFM studies.

The question, whether there is a thickness induced change of the film morphology from amorphous tantalum to cyrstalline modifications for films thicker than 100 nm with negligible oxygen contents, must be left open at this moment and will be adressed later [213].

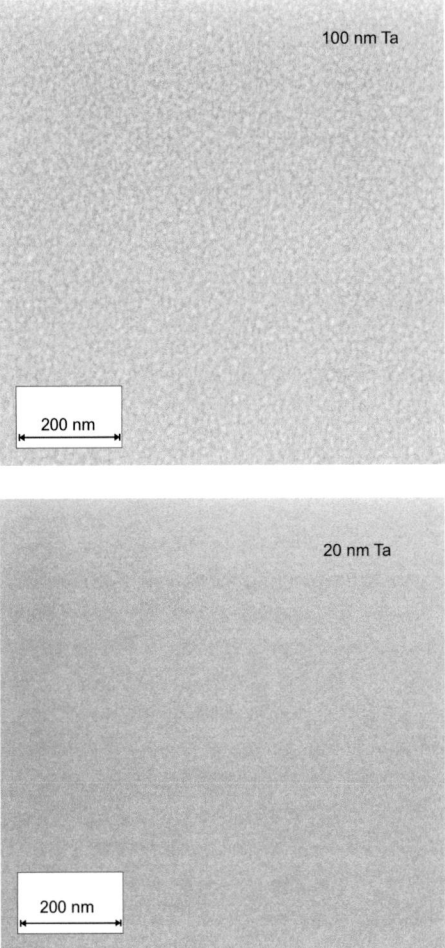

Figure 4.23: Scanning electron microscopy images of 100 nm and 20 nm thick tantalum films. Magnification: $2 \cdot 10^5$.

4.4 Preparation and properties of thin amorphous tantalum films

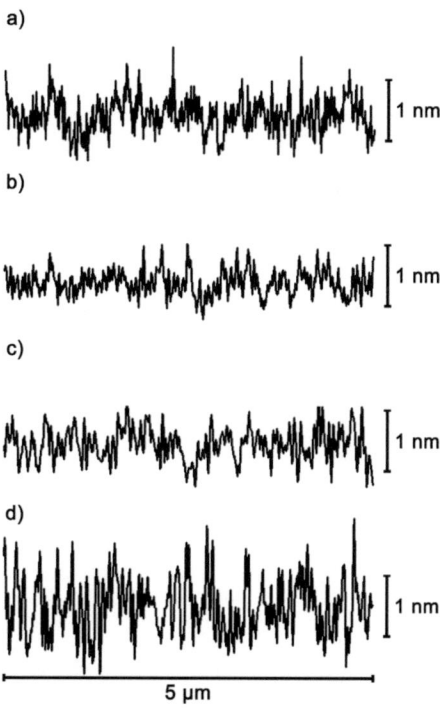

Figure 4.24: AFM linescans obtained for 10, 20, 40 and 100 nm thick tantalum films (graph a) - graph d)).

Characterization of the tantalum film by AFM

To crosscheck the absence of an overlaid structure in the SEM images, we performed additionally AFM studies for 10, 20, 40 and 100 nm thick films. AFM is a suitable method since it can detect edge structures > 5 nm [214]. Therefore one could clearly determine grains or other e.g. 100 - 200 nm large overlaid structures.

As suggested by the SEM images we can detect only small roughness values. The linescans a-c in fig. 4.24 show roughnesses that lie in the detection limit of ≈ 2 nm [214]. Only the scan for the 100 nm sample shows a roughness value slightly larger than 2 nm. It shoud be noted that the roughness shown in our AFM line scans of around 2 nm is clearly smaller than the values reached for sputtered tantalum films on glass of around 10 nm [184].

4 Measurements, results and discussion

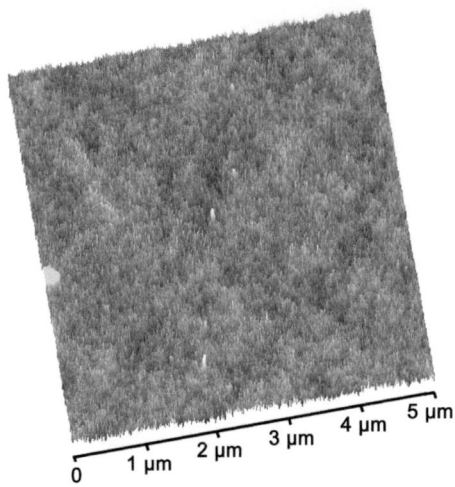

Figure 4.25: Topography of an 20 nm thick tantalum film with a roughness of ≈ 2 nm.

In fig. 4.25 an AFM-topography of the tantalum is shown for the 20 nm thick sample with ≈ 2 nm roughness. It confirms the absence of any overlayed macroscopic structure on a 5 μm scale.

Apparently the AFM and SEM studies point to a very smooth structure (for the 10 nm - 40 nm samples) with only a small structure change for the 100 nm sample.

However, both methods cannot distinguish between nanocrystalline structures (diameter < 2 nm) and a completely amorphous film. Some further hints pointing to a prevalent amorphous structure will be given by resistivity measurements in a later section.

For a crosscheck we performed X-ray-diffraction measurements in $\theta - 2 \cdot \theta$ geometry. The scan did not show any peaks, only a broad maximum at around $2\theta = 25^0$ could be seen. This also points to a completely disordered structure.

The determination of the tantalum film thicknesses by AFM requires a special preparation of the film, since an atomic force microscope needs localized sharp edges (scan range $\approx 10\,\mu$m). The scratching of thin tantalum films with metals, for example steel needles, does not harm the metal with modest pressure. With higher pressures the metal can be removed, but damages in the underlying glass cannot be excluded.

4.4 Preparation and properties of thin amorphous tantalum films

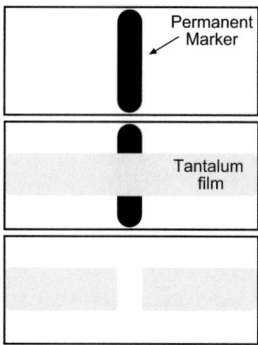

Figure 4.26: **Upper graph:** Scheme of the permanent marker line on the glass substrate. **Middle graph:** Tantalum film evaporated above the marker on the glass substrate. **Bottom graph:** After removing the marker line with an ethanol wetted tissue, this part of the tantalum film can be peeled off.

Therefore a different procedure for the formation of sharp edges in tantalum films has to be applied. We use a commercial permanent marker to produce a thin line on the glass substrate, which acts as a spacer layer between tantalum and glass in the course of the deposition process. After the deposition procedure, the polymer of the marker with the weakly bound tantalum layer on top can be easily wiped off with an ethanol wetted tissue (see fig. 4.26). This procedure produces an extremely sharp edge, comparable to a breaking edge in a crystal without modifying the other parts of the sample (see fig. 4.27). All thickness values used in the present section are gauged by AFM height measurements accomplished on edges of the tantalum films prepared by this method.

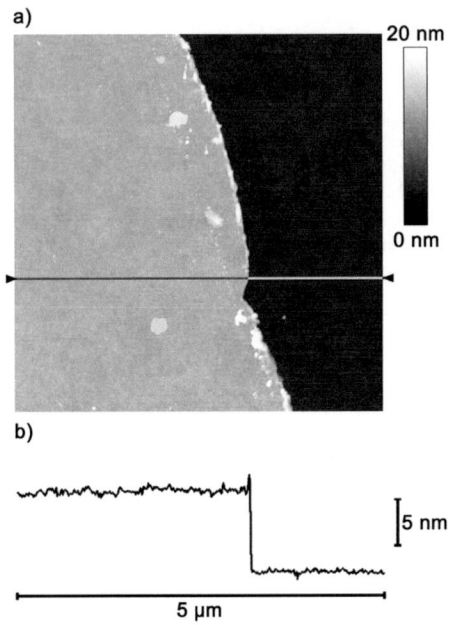

Figure 4.27: **Upper graph**: Topography across the sharp edge of the film. **Lower graph**: AFM linescan across the sharp edge of the film.

4.4.2.3 Growth of tantalum films on silicon

In addition we performed SEM and X-ray-diffraction measurements of tantalum thin films deposited on n-doped Si(111) pieces. Both experimental methods yield similar results as for the tantalum–glass system. The SEM did indicate any structures. In the XRD no line besides the substrate induced line at $2\theta = 27°$ could be observed.

Therefore I can exclude that the growth of amorphous tantalum films is a specific property of the tantalum–glass interface.

4.4.2.4 Electrochemical characterization of the tantalum film

Thin metal films can be characterized with respect to their homogeneity and mechanical stability by an electrochemical oxidation procedure. Single crystals and smooth homogeneous tantalum films show after the onset potential a constant oxidation current when the electrochemical potential is increased linearly with time (see section 4.3 and ref. [215, 216]). An example for a freshly prepared tantalum film is presented in the upper graph of fig. 4.28. An potential $E_{initial}$ of -1.2 V is initially applied. At this potential no chemical reaction occurs. Then the potential is increased with a rate of 0.1 V/s up to a final value of $E_{final} = 1.8$ V. At around -0.3 V (denoted as $E_{1/2}$ - half wave potential) a considerable current increase is monitored, followed by a constant current density of around 500 $\mu A/cm^2$. When the electrode potential is held at the final value of 1.8 V, the current drops exponentially to zero within a few seconds. The potential driven chemical reaction occuring on the tantalum film is a pure oxidation. Parallel processes as corrosion can be neglected since the oxide has a negligible solubility in the sodium acetate buffer electrolyte used here [88]. Under these conditions it is known that the final thickness value of the built oxide layer scales linearly with E_{final} [88].

The middle graph in fig. 4.28 shows a cyclovoltammogram of a tantalum film, which was exposed to ambient conditions for 10 hours. One can see the delayed current increase (corresponding to a shift of $E_{1/2}$ by $+0.35\ V$) which is due to a oxidation of the tantalum by ambient oxygen prior to the experiment. The mentioned shift corresponds to a thickness increase of 0.7 nm [121]. Further increasing of the potential leads to a constant current density of around 500 $\mu A/cm^2$ as for the sample shown in the upper graph. When the final value of 1.8 V is reached and kept constant, the current drops exponentially as function of time similar to the experiment mentioned above.

4 Measurements, results and discussion

In the bottom graph a cyclovoltammogram of a site on the tantalum film with a dust enclosure is shown. Up to a potential of -0.2 V the experimental curve looks quite similar to the curves in the upper and middle graph. But at a potential of $E \approx 1.0$ V a considerable current increase is monitored, followed by a second one at 1.5 V. When the potential is held constant at -1.8 V, the current does not simply fall exponentially to zero as in the previous experiments, but exhibits several sudden current spikes. After the end of the oxidation this film shows mechanical cracks whereas the samples with smooth cyclovoltammograms do not exhibit any visible modifications.

The reason for this oxidation behaviour can be seen in the strong mechanical tension of the oxidized metal which is due to the volume increase [217, 218]. Discontinuities as dust enclosures then tend to break up. The break up will supply new, unoxidized metal to the electrolyte. This is most probably the reason for the sudden current increases in the bottom graph of fig. 4.28.

XPS sputter profiles were taken from the anodically oxidized samples. Incorporated cations (sodium in this case) could not be found in the oxide layer. The oxygen contents of the tantalum film itself was found to be below 1 % [212].

4.4.3 Electrical resistivity of thin tantalum films

Furthermore, the tantalum films were characterized by measurements of their electric resistivity. The set-up for the resistivity measurements in this section (using 2 mm wide and 2 cm long stripes) was described in section 4.3. In the upper graph of figure 4.29 the specific resistivity ρ of the tantalum film is plotted as a function of the film thickness d. The values are calculated with the resistance R of the samples by

$$\rho = R \cdot \frac{d \cdot b}{l} \qquad (4.11)$$

where l is the length and b the width of the sample ($l = 20$ mm, $b = 2$ mm). This is only meaningful if the topography of the film does not change significantly with the film thickness [219, 220]. However, this seems to be at least partly fulfilled for these samples as presented in figure 4.24 and 4.23. The observed values lie in the range of some $100\,\mu\Omega \cdot$ cm (red crosses in upper graph of figure 4.29) and are more than 10 times larger than the value for bulk tantalum at room temperature ($\rho = 13.4\,\mu\Omega \cdot$ cm [221], see black line (c) in upper graph of figure 4.29).

The high resistivity values observed here are even higher than values for liquid

4.4 Preparation and properties of thin amorphous tantalum films

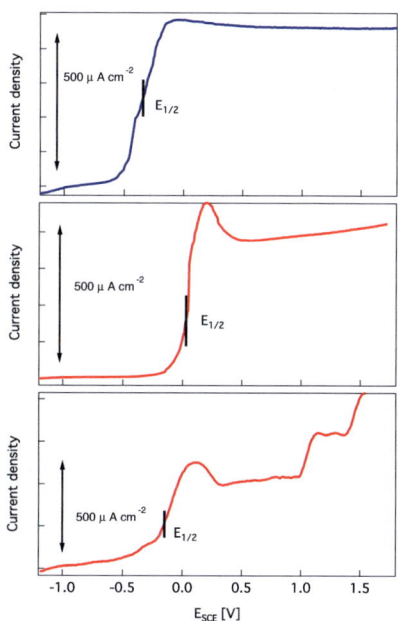

Figure 4.28: Cyclovoltammograms of tantalum films ($E_{\text{initial}} = -1.2$ V, $E_{\text{final}} = +1.8$ V, $dE/dt = 0.1$ V/s). Straight lines denote the half wave potential $E_{1/2}$. **Upper graph:** Freshly prepared film. **Middle graph:** Tantalum film exposed to ambient conditions for 10 h. **Bottom graph:** Tantalum film which includes a dust enclosure.

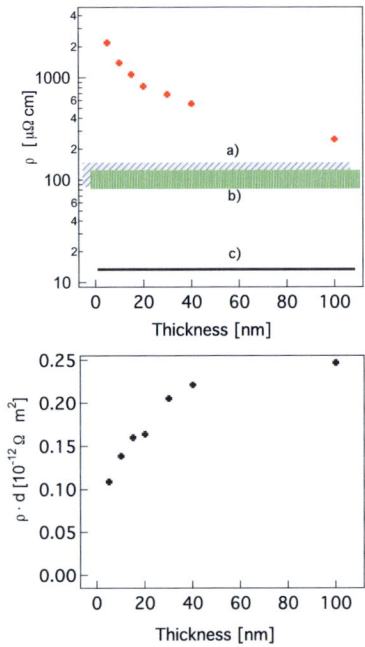

Figure 4.29: **Upper graph:** Specific resistivity ρ (red crosses) of the tantalum films as function of the film thickness. Values of ρ for molten tantalum (boxes a) and b)) according to [222] and [223], respectively. Horizontal line c) denotes the value for bulk tantalum [221]. **Lower graph:** Product of specific resistivity ρ and film thickness d as function of d.

tantalum. For liquid phase tantalum a specific resistivity of $80 - 150\,\mu\Omega\cdot\text{cm}$ [222] respectively $85 - 120\,\mu\Omega\cdot\text{cm}$ [223] is observed. Both value ranges can be seen as horizontal boxes (a) and b) in the upper graph of figure 4.29). In summary these results point to a structure of the tantalum films having even a higher disorder than a liquid.

To exclude surface effects or size effects due to the limited thickness of the films, the resistivity data are analysed with the classical Fuchs-Sondheimer approximation [203, 224] according to

$$\rho \cdot d = \rho_\infty \cdot d + \frac{3}{8}\rho_\infty l_\infty \cdot (1-p), \qquad (4.12)$$

where ρ_∞ is the resistivity of an infinitely thick film with the same structure as the investigated film. l_∞ is the analogously defined mean free path and p the specularity parameter (fraction of electrons being mirror-like reflected at the film interfaces). A $\rho \cdot d$ vs. d plot is shown in the lower graph of figure 4.29. One can clearly see that there is no linear behavior. Thus, an evaluation with the Fuchs-Sondheimer theory seems to be ruled out.

If we disregard the data point at 100 nm, one could be tempted to try a linear regression with points between 5 nm and 40 nm. Thus, an evaluation according to Fuchs-Sondheimer could be tested. The slope of the regression would lead to a value of ρ_∞ of $3.12 \cdot 10^{-4}\,\Omega\cdot\text{cm}$.

The second term in the sum of equation 4.12 must then be equal to the ordinate axis intercept. If we assume $p = 0$ we would get a minimum value for l_∞ of 68 nm. This value can be interpreted as a scattering time τ by

$$\tau = \frac{l_\infty}{v_F} \qquad (4.13)$$

where v_F is the Fermi velocity (for Ta $v_F = 1.79 \cdot 10^6\,\frac{\text{m}}{\text{s}}$ [225]).

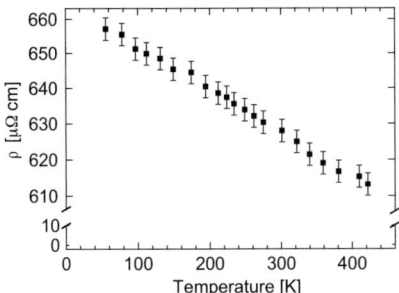

Figure 4.30: Temperature dependence of the specific resistivity ρ of a 40 nm thick tantalum film.

According to Drude's resistivity approach:

$$\rho = \frac{m}{n \cdot e^2 \cdot \tau} \quad (4.14)$$

one can then evaluate a value for ρ. This theoretical value derived by assuming the validity of the Fuchs-Sondheimer approach is $0.7\,\mu\Omega \cdot$ cm. This value is much too low compared to the experimental value of some $100\,\mu\Omega \cdot$ cm. Thus an evaluation within the Fuchs-Sondheimer theory is not appropriate for our tantalum films.

A further indicator for a disordered structure is the temperature coefficient of the resistivity (TCR). This value, $\alpha(\rho)$, is commonly evaluated by [209]:

$$\alpha(\rho) = \frac{1}{\rho} \cdot \frac{d\rho}{dT}. \quad (4.15)$$

α can be expected to be positive for crystalline films, whereas it should be negative for amorphous tantalum films [199]. Following the theory of Mooij and Tsuei [208, 209], one can expect high negative $\alpha(\rho)$ values for specific resistivities $\rho > 500\,\mu\Omega \cdot$ cm. For that reason I present the temperature dependent resistivity of a 40 nm thick tantalum film (ρ at 300 K $\approx 630\,\mu\Omega \cdot$ cm) in figure 4.30. This sample shows a reproducible linear increase of the resistivity when cooled from 450 K to 60 K. From the slope one obtains negative TCR values (typically $-2.3 \cdot 10^{-4}\,1/K$ and $-2.1 \cdot 10^{-4}\,1/K$ for 15 and 40 nm thick films) according to equation 4.15.

These results are in good agreement with the data collected by Tsuei [209]. For

15 and 40 nm thick films (specific resistivities of 1066 $\mu\Omega \cdot$ cm and 630 $\mu\Omega \cdot$ cm, see figure 4.29) the values of $\alpha(\rho)$ are definitely negative (similar to the results for pulsed laser deposited tantalum [199]). This agreement and the homogeneous smooth structure of the films lets me assign the negative α values to an intrinsic property of the electron transport in amorphous tantalum.

4.4.4 Conclusion

In the present section I showed how thin amorphous tantalum films can be prepared using small e-beam evaporators. With a distance of $\approx$ 20 cm between source and sample, homogeneous films with thickness variations below 1 % can be achieved.

The SEM images, AFM studies and the resistivity measurements point to a completely amorphous and smooth structure of the thin tantalum films (grown on glass and silicon substrates). This result is supported by the negative TCR (temperature coefficient of resistivity) of $\approx -2 \cdot 10^{-4}$ 1/K of the thin films which is in agreement with literature values for highly disordered metals.

Despite their amorphous character the adhesion of tantalum films on glass slides is strong. Therefore the films cannot be peeled off easily. The fixing of the film thickness by AFM measurements across a sharp defined tantalum edge is possible with a classical 'lift off' procedure. The surface of the films was found to be quite smooth (roughness $\approx$ 2 nm). Due to their high hardness and adhesive power on the glass, the tantalum films can be reproducibly modified by an electrochemical oxidation procedure without any cracks or mechanical damages. Selectable oxide thicknesses up to 4 nm can be accomplished by applying oxidation potentials up to 2 V.

The smooth and continuous structure of the thin films allows their application in the manufacturing of heterojunctions. These can be metal–insulator–metal structures [7,9,85] as well as the application in stacked metals [226]. Furthermore very thin and stiff cover layers (also for use in STM) may be formed with these films. With respect to biocompatibility elemental amorphous tantalum might be of advantage over alloys [227].

4 Measurements, results and discussion

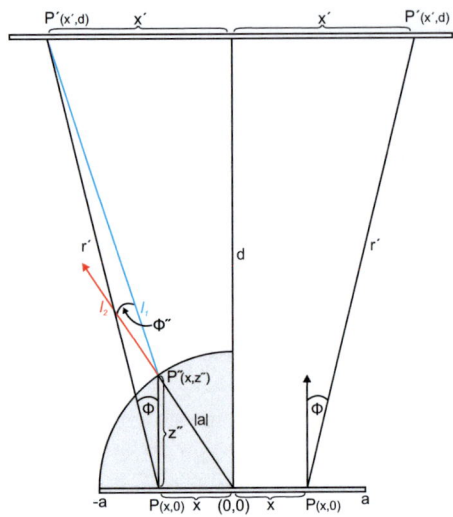

Figure 4.31: Comparison of the geometries for evaporation from a planar evaporation source (**right side**) and from a hemispherical evaporation source (**left side**).

Appendix:

Theoretical description of film thickness distribution: Comparison of plane and hemispherical sources

In the appendix I want to give a short description of the film thickness distribution caused by evaporation from ample sources. Fig. 4.31 shows on the right hand side the geometry for evaporation from a planar evaporation source of width $2\,a$, whereas the left hand side shows the geometry for hemispherical sources like a molten rod with radius a.

The target plane is in a distance d from the source plane. On the source plane the coordinates are unprime notations in the case of a planar source, or double prime notations when the considered point of the evaporation is on the hemisphere's surface. On the target plane the coordinates are in prime notations. The differential mass flux $d\Phi_{P'}/dx$ on a point $P'(x',d)$ caused by evaporation from the point $P(x,0)$ is proportional to $(r')^{-2}$ and $cos(\phi)$ where $r' = \sqrt{(x'-x)^2 + d^2}$ [228–230]. With $cos(\phi) = d/r'$ one obtains:

82

4.4 Preparation and properties of thin amorphous tantalum films

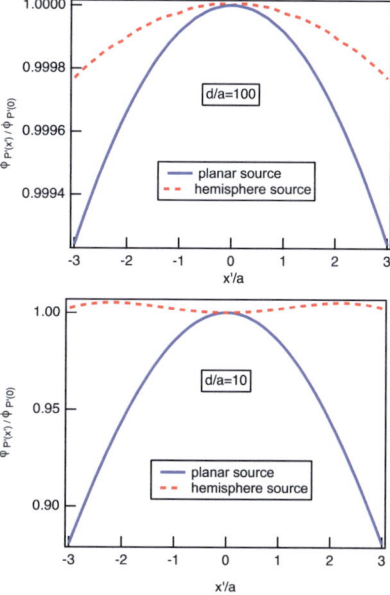

Figure 4.32: Calculated thickness distribution for a hemispherical evaporation source and for a planar evaporation source. **Upper graph:** Ratio of distance to radius of the evaporation source $d/a = 100$. **Lower graph:** $d/a = 10$.

$$\begin{aligned}\frac{d\Phi_{P'}}{dx} &= C \cdot \frac{1}{r'^2} \cdot \cos(\phi) \\ &= C \cdot \frac{1}{(x'-x)^2 + d^2} \cdot \frac{d}{\sqrt{(x'-x)^2 + d^2}}\end{aligned} \quad (4.16)$$

Considering a planar evaporation source of the width $2a$, one obtains the total mass flux $\Phi_{P'}$ on P' caused by the whole evaporation source as:

$$\Phi_{P'} = \frac{C}{2a} \int_{-a}^{a} \frac{d}{((x'-x)^2 + d^2)^{3/2}} dx \quad (4.17)$$

In the case of evaporation from a molten rod, having a hemisphere with the radius a on top, one has to consider a simple transformation of coordinates. The

4 Measurements, results and discussion

point $P(x, z = 0)$ on the source plane moves to the point $P''(x, z'')$ of the hemisphere's surface. So the transformations for $P \rightarrow P''$ are

$$
\begin{aligned}
x &\rightarrow x'' & x &= x'' \\
z = 0 &\rightarrow z'' & z'' &= \sqrt{a^2 - x^2}
\end{aligned}
$$

The total mass flux at $P(x', d)$ can be formulated in this case analogously to equation 4.16:

$$\Phi_{p'} = \frac{C}{2a} \int_{-a}^{a} \frac{\cos(\phi'')}{((x' - x)^2 + (d - z'')^2)} dx \qquad (4.18)$$

One has to consider that the cosine like evaporation profile on P'' must be calculated with respect to the surface normal in P''. Hence, ϕ in equation 4.16 has to be transformed in ϕ'' in equation 4.18.

ϕ'' is the included angle of the straight lines l_1 and l_2 (see figure 4.31). l_2 is the surface normal in P'', l_1 is the straight line $P' - P''$. The slopes of the two lines are $(z'' - d)/(x - x')$ for l_1 and z''/x for l_2. The included angle can then be calculated to:

$$\phi'' = \frac{\pi}{2} - \left(\arctan(\frac{z'' - d}{x - x'}) + \arctan(\frac{z''}{x}) \right) \qquad (4.19)$$

Replacing ϕ'' in equation 4.18 by equation 4.19 allows the direct comparison of the thickness distribution on a flat sample caused by a molten rod with one caused by a planar evaporation source using equation 4.17.

In figure 4.32 the results for planar and hemispherical sources are compared for targets close to (lower graph) and far away (upper graph) from the evaporation source (for example $a = 1.5$ mm and $d = 15$ mm or 150 mm respectively). The constant C in the previous equations is set to 1, because it does change the thickness distribution. The range of the x-axis is chosen to be three times the radius of the source. The mass flux is plotted relative to the value on the center axis at $x = 0$. In the upper graph for $d/a = 100$ one can see that the planar and the hemispherical source show a negligible thickness variation smaller than 1 %. This changes for distances close to the source (lower graph). For $d/a = 10$ the thickness distribution gets much more pronounced for planar sources. A decrease of up to 12 % is calculated for the planar source. In contrast, the hemispherical source shows a much smaller variation in the thickness distribution. Additionally, the hemispherical source exhibits pronounced maxima at $x \approx 2.2 \cdot a$. This leads to the sometimes observed corona like thickness distributions. The maxima and the corresponding local minimum at $x = 0$ are due to the projection of the hemisphere's cosine distributions on the flat target plane.

In summary, the hemispherical sources produce more homogeneous thickness distributions for close as well as distant targets.

4.5 Thin tantalum films on crystalline silicon – a metallic glass[3]

4.5.1 Introduction

Recently Wu et al. [232] and Errandonea [233] raised the question whether metals can be a liquid glass. This is not only a matter of debate with significance for high pressure melting studies where tantalum stands for a metal for which each atom experiences the same interatomic potential, but as well for thin film studies since thin tantalum films may play a key role in future micro- and nanoelectronic sensors [187].

Hence, the preparation of thin tantalum films is still of continuing interest despite the vast literature accumulated in the last four decades (see [188], [196] and references therein). In a previous section (4.4) I showed how thin, smooth, and continuous amorphous tantalum films can be prepared on glass or silicon substrates. Inspired by Wu et al. [232] however I try to go the opposite way of preparation: from the liquid–like glass, i.e. an amorphous structure of thin tantalum films, to a crystalline structure, in this case $TaSi_2$. Therefore thin amorphous tantalum films on a crystalline (silicon) and on an amorphous (glass) substrate were prepared. Then I induced a phase transition to a crystalline metal out of an amorphous metal and proved the transformation of the amorphous tantalum by recording the temperature coefficient of the resistivity (TCR) (compare scheme in figure 4.33). Our data sheds new light onto the correlation between the specific electrical resistivity and the TCR suggested by Mooij [208].

4.5.2 Experiment and Discussion

4.5.2.1 Resistivity

Amorphous tantalum films were prepared on microscope glass slides and on a 525 μm thick silicon n-Si(111) 7.5 $\Omega \cdot$ cm ($5 \cdot 10^{14}$ cm^{-3} phosphorus doped) piece ($2 \cdot 1$ cm^2), kept at room temperature, using a commercial e-beam evaporator (see section 4.4) under Ultra High Vacuum conditions. A very slow deposition rate, 0.006 nm/s, was used in contrast to many other experiments [196,234]. Tantalum was deposited from the atom beam only on an area of $1.4 \cdot 0.95$ cm^2 of the silicon substrate. The silicon was contacted on the sides using tantalum clamps allowing

---
[3]This section was slightly changed from its published form in phys. stat. solidi RRL 5, 68 (2011) [231].

4.5 Thin tantalum films on crystalline silicon – a metallic glass

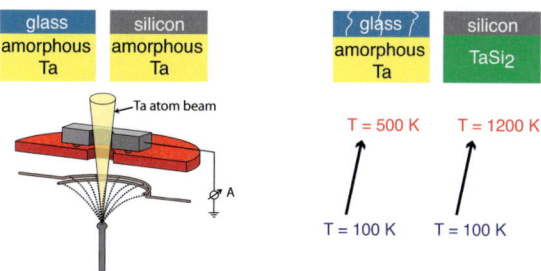

Figure 4.33: Schematic drawing of the evaporation set-up on either glass or silicon samples. Scheme of annealing effects.

us to heat the sample directly by applying a voltage (analogously to the set-up shown in figure 4.5). The glass sample was heated indirectly utilizing a tungsten filament mounted on the sample holder. Additionally, the sample holder was connected to a helium cryostat. The temperature was monitored using a Pt1000 sensor and a pyrometer. The tantalum films were connected on the front side to a potentiostat to measure their resistance. For this purpose a constant voltage of 0.01 V was applied and the resulting current was recorded.

The temperature dependence of the film's resistance was always recorded over the temperature range from 148 K to 290 K to allow a comparison for all temperatures reached in the experiment, analogous to ref. [234]. These temperature curves were measured during the cooling phase after reaching the higher temperatures to avoid any artifacts caused by the direct heating of the underlying silicon in the warming period.

Figure 4.34 shows the temperature dependence of the resistivity of a 40 nm thick tantalum film on silicon. The specific resistivity ρ, $\rho = R \cdot \frac{d \cdot b}{l}$, of the tantalum film is depicted, where d is the film thickness, l its length and b its width ($d = 40$ nm, $l = 1.4$ cm, $b = 0.95$ cm) and R the recorded resistance. The evaluation of ρ is sensible since I showed in a previous section (4.4) that the films are very smooth in the considered thickness range.

For the lowest temperature in my experiment I, obtained that $\rho = 2000 \, \mu\Omega \cdot \text{cm}$. Up to 180 K the resistivity decreases almost linearly with temperature. The slope of the temperature dependence (TCR) $\alpha(\rho)$ is evaluated by

$$\alpha(\rho) = \frac{1}{\rho} \cdot \frac{d\rho}{dT} \qquad (4.20)$$

4 Measurements, results and discussion

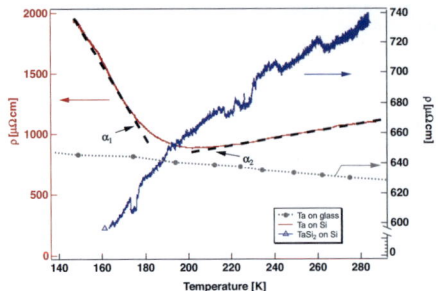

Figure 4.34: Temperature dependence of the specific electrical resistivity of a 40 nm thick tantalum film on glass (grey dotted curve with bullets) and silicon (red curve) substrates. The blue curve with triangles denotes the data for TaSi$_2$. The different slopes of the resistivity curve of the tantalum film on silicon are observed and marked with α_1 and α_2.

yielding $\alpha_1 = -1.62 \cdot 10^{-2}\,\text{K}^{-1}$. Extrapolating the resistivity to $T = 0$ K yields $\rho(T = 0\,\text{K}) = 6300\,\mu\Omega\cdot\text{cm}$. Such a high value is commonly attributed to scattering induced electron localization in an amorphous metal and can be estimated by (compare section 4.4):

$$\rho(T = 0\,\text{K}) = \frac{3\pi^2 \hbar}{e^2} \frac{l_0}{(k_F \cdot l_0)^2}, \qquad (4.21)$$

where k_F is the Fermi vector of the metal (E_F = 9.5 eV for tantalum corresponding to $k_F = 1.58 \cdot 10^{10}\,\text{m}^{-1}$).

We assume that the product $k_F \cdot l_0$ (l_0 is the scattering length) is between 2.3 and 2.5, since the local minimum of the resistivity's temperature dependence at $T \approx 200$ K coincides quite well with the data calculated in ref. [235]. This suggests 4.3 Å - 5.1 Å for the scattering length l_0.

Based on my previous studies (section 4.4), I assign the decrease of the resistivity between 148 K and 180 K to the films amorphous character, which is the reason for a negative value of the temperature coefficient of the resistivity [209]. But, the value of $\alpha_1 = -1.62 \cdot 10^{-2}\,\text{K}^{-1}$ is two orders of magnitude larger than the literature values compiled by Tsuei [209]. At first one might suspect an influence from the underlying silicon substrate, since highly doped silicon also shows a negative temperature coefficient. But for dopant concentrations as used here ($5 \cdot 10^{14}\,\text{cm}^{-3}$) only positive temperature coefficients are expected [236]. Hence, I rule out a contribution from the underlying substrate.

4.5 Thin tantalum films on crystalline silicon – a metallic glass

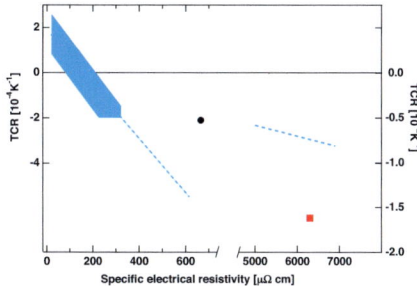

Figure 4.35: TCR vs. specific electrical resistivity (Blue shaded polygon indicates the region where the original Mooij data points can be found [208]). The dotted blue line represents the mean slope of the data in the polygon (shown for both resistivity ranges). The black bullet represents the value for tantalum films on glass presented in section 4.4. Red square indicates the value for tantalum films on silicon extrapolated to 0 K from figure 4.34.

Mooij [208] observed that the data in the literature fall into a particular region of a TCR vs. ρ graph (blue shaded polygon in figure 4.35). He suggested a linear correlation between these two quantities (dotted blue line) which was later corroborated using an extended data set and further supported with numerical modeling by Tsuei [209]. Our TCR value for tantalum on glass is larger than suggested by this relationship. The value for tantalum on silicon is smaller than suggested by this trend line. However, these deviations are not in conflict with the general correlation when taking into account the large scatter of the individual data points which underpin the correlation. Thus, our data support the correlation between α and ρ established in the literature and suggest that it can even be extrapolated to considerably higher resistivity values.

Above 200 K the tantalum film on silicon shows a positive TCR value, $\alpha_2 = +2.59 \cdot 10^{-3}\,\mathrm{K}^{-1}$, indicating that we are now in the usual Boltzmann transport regime where electron phonon scattering dominates.

It should be pointed out, that the morphology does not change (as observed earlier for crystalline tantalum phases [188, 234]) when varying the temperature from 148 K to 290 K. The local minimum at 200 K is found reproducibly and the $\alpha_{1,2}$ values are constant within some percent.

The chemical composition of the amorphous tantalum films on silicon can be checked by an annealing experiment. Following the literature [237, 238], high

4 Measurements, results and discussion

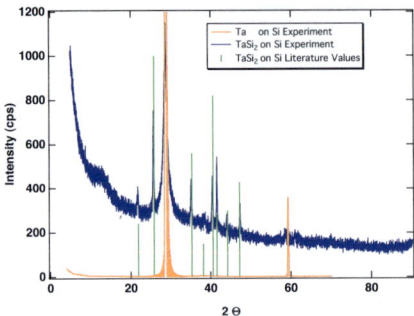

Figure 4.36: XRD diffraction pattern of tantalum on silicon (orange curve), TaSi$_2$ on silicon (blue curve) and literature values (green curve) for TaSi$_2$ on silicon [240].

temperatures, up to 1200 K, were applied by heating the silicon substrate under Ultra High Vacuum conditions. The film is irreversibly transformed and exhibits afterwards a small positive TCR value of $+1.57 \cdot 10^{-3}\,\text{K}^{-1}$ (see figure 4.34) which deviates just by a factor of two from the literature value of $+3.32 \cdot 10^{-3}\,\text{K}^{-1}$ for TaSi$_2$ bulk samples [239].

No structural changes are observed when annealing the amorphous tantalum films prepared on glass up to 500 K. At higher temperatures the glass begins to melt and structural changes in the glass substrate are observable, leading to a destruction of the sample (see figure 4.33).

4.5.2.2 X-ray diffraction

To study the structure of the tantalum films on silicon we recorded XRD diffraction patterns silicon. No lines besides the ones due the silicon substrate at $2\theta = 27°$ and $2\theta = 59°$ appear (see figure 4.36). It should be mentioned that these XRD diffraction pattern do not change with sample temperature from 148 K to 300 K. After annealing the sample to 1200 K, we find a clear signature of the formation of crystalline TaSi$_2$. The observed then maxima in XRD are in agreement with the literature values for TaSi$_2$ [240] as indicated in figure 4.36. Additionally, the absence of other diffraction patterns confirms that the sample is clean and no side reaction of tantalum with other species than silicon occurs. It should be pointed out, that TaSi$_2$ can be directly formed, when energetic Ta-ions impinge on a silicon surface [241]. This process can be ruled out in the experiments since the e-beam evaporator with sample source distance of 20 cm, as used in the experiments, is a pure atom beam with small kinetic energies.

4.5.3 Conclusion

Amorphous tantalum films on silicon exhibit an extraordinary high resistivity value ρ of 6300 $\mu\Omega \cdot$ cm (extrapolated to 0 K) and a high negative TCR value α. The combination of α and ρ is located on the extrapolation of Mooij's α vs. ρ plot. Thus, the formation of metallic glassy tantalum on silicon is possible under Ultra High Vacuum conditions.

4.6 Charge transport through thin amorphous TiOx and TaOx layers[4]

4.6.1 Introduction

Heterosystems with thin oxide layers are interesting for a variety of applications such as thin film capacitors [156, 242, 243], chemicurrent detectors [7, 9] and electron sources in spin polarized tunneling [244]. In this section a comparative investigation of charge transport induced by different methods through thin oxide films is shown. Charges are driven through the oxide by: i) application of a sensor voltage ii) illumination of the samples with monochromatic light at variable photon energies, leading to spectra of internal photoemission, iii) non-adiabatic chemical surface reactions. All three methods are combined to allow us to shed light on the nature of excited carriers transport through heterosystems [14]. For aluminum and tantalum oxide heterosystems it was found, that they form a high pass filter for excited electrons and holes (defect electrons) [7, 9, 14]. Since the height of the internal barriers for electrons and holes can be modified by an applied bias voltage, it became possible to characterize the spectra of excited charge carriers with energies below the vacuum barrier [107]. Bias voltages of up to 1 V could be applied to the system which had an internal barrier height of 3 eV, without changing the dielectric properties of the samples.

To increase the detection efficiency for excited electrons travelling from the surface of a metal towards the bulk, one can either decrease the thickness of the top metal film of a heterosystem or one can reduce the height of the internal barrier. But with a decrease of the internal barrier the method of applying a bias voltage for tuning this barrier might become problematic, since tunnel currents become larger with lower barrier heights. Additionally, the influence of midgap states, impurities and remanent changes of the internal barrier might become more relevant for lower barriers [149, 245]. These problems are adressed in the present work.

A comparison of metal/insulator/metal heterojunctions is presented with potentiostatically formed titanium and tantalum oxide as interjacent insulating layer. Titanium and tantalum oxide were chosen since both have bandgaps smaller than 4.5 eV [246]. For amorphous tantalum oxide values of 4.2 eV are typical [247, 248] whereas crystalline samples show values of 3.9-4.5 eV [248, 249]. Bulk

---

[4]This section is slightly modified from its published form in J. Electrochem. Soc. 158, P65 (2011) [83].

titanium oxide values are for rutile 3.03 eV and for anatase 3.2 eV. For polycrystalline titanium oxide films 3.34 eV - 3.39 eV are typical values [250]. The electrochemical oxide formation is also well known in the literature [121, 215, 251–253]. The electrochemical preparation is described in section 4.6.2.1 based on potentiodynamic growth in acetate solution, followed by the preparation of the gold top electrodes in vacuum as described in 4.6.2.2. The gold top electrodes seem to form a more open structure on titanium oxide than on tantalum oxide. The growth is monitored by in situ resistivity measurements during deposition of the metal film.

4.6.2 Sample preparation

4.6.2.1 Electrochemical oxidation

As heterojunctions I employed thin film Ta–TaOx–Au and Ti–TiO–Au sandwich structures. The tantalum bottom electrode (2 mm · 20 mm) was prepared by a mini e-beam evaporator under Ultra High Vacuum conditions (UHV) as discussed previously in sections 4.4 and 4.5 whereas the titanium bottom electrode (2 mm · 20 mm) was evaporated using a bent beam evaporator under UHV conditions [254].

The insulating oxide layers were prepared by a localized electrochemical oxidation procedure in an electrolytic droplet cell (see section 4.3 and references [85, 86]). In order to minimize a parallel corrosion processes during the oxidation [87], a sodium acetate buffer electrolyte and an ammonium acetate buffer electrolyte (0.9 mol/L) were used for Ta and Ti, respectively. While an initial potential $E_i = -1.0$ V is applied no chemical reaction takes place on Ta. For the Ti samples a slight increase of the negative current can be seen already for electrode potentials $E_{SCE} < -0.8$ V. This can be attributed to a beginning hydrogen evolution at the Ti sample. The anodic oxidation is then initiated by increasing the potential with a constant rate $dE_{SCE}/dt = 0.1$ V/s up to a final value E_{final}. In the present work, tantalum films oxidized at $E_f = 2$ V as well as titanium films oxidized at $E_{final} = 2$ V and $E_{final} = 3$ V were used. The corresponding cyclovoltammograms, recorded during the oxidation of the Ta and Ti films, covered by a thin native oxide due to previous exposure to ambient air, are shown in Figure 4.37. In all three cases, a similar behavior can be observed: up to a certain positive potential $E_I \approx 0$ the current density remains low, then it increases and goes through a maximum, and finally reaches a constant plateau value of some 10 μA until E_{final} is reached. At a constant value of E_{final}, the current drops expo-

4 Measurements, results and discussion

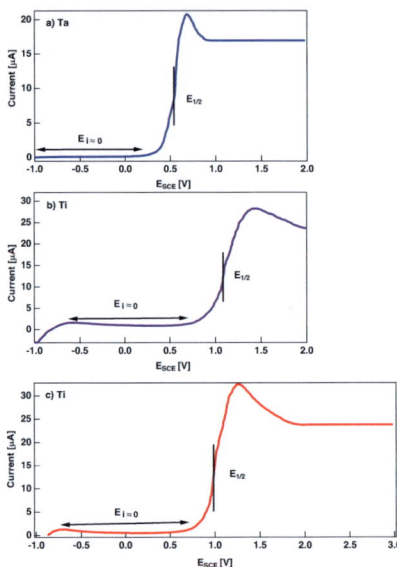

Figure 4.37: Typical cyclovoltammograms ($E_i = -1.0\,V$ and $E_f = 2.0 - 3.0\,V$) recorded during the localized electrochemical oxidation in an electrochemical droplet cell of a tantalum (**upper graph**), and a titanium (**middle and lower graph**) substrate over an area $A_{ox} \approx 4\,\mathrm{mm}^2$, respectively, where $E_{1/2}$ denotes the half wave potential.

nentially to zero within a few seconds. The so-called half wave potential $E_{1/2}$ is usually introduced as a characteristic quantity describing the oxidation process. It is defined as the potential at which the current density is half of its plateau value. For tantalum, $E_{1/2}$ is about +0.5 V, while for titanium it is around +1.0 V. The small difference in $E_{1/2}$ between the two titanium samples is caused by a thicker oxide due to a longer exposition to ambient conditions after deposition in case of the sample for which the middle graph of figure 4.37 was obtained.

One advantage of this oxidation procedure is the fact that it allows to control the oxide thickness z_{ox} by varying E_{final}, since the relation between the two quantities is approximately linear [88]. From this linear dependence, the so-called formation factor $dz_{ox}/dE_{\mathrm{final}}$ was found to be in the range 1.3–2.4 nm/V for TaOx [121] and 1.3–3.3 nm/V for TiOx [121]. For our sensors, we derived, from XPS sputter profiles, film formation factors of about 2.2 and 2.4 nm/V for the tantalum and the titanium oxide, respectively. This calibration factor gives

4.6 Charge transport through thin amorphous TiOx and TaOx layers

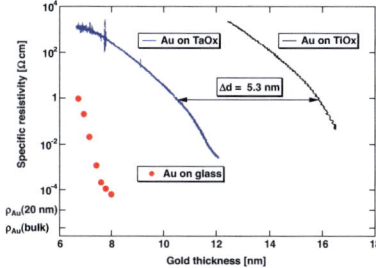

Figure 4.38: Resistivity of the Au film evaporated on tantalum oxide, titanium oxide and on glass (measured at 300 K). $\rho_{\text{bulk}}(\text{Au}) = 2.2 \cdot 10^{-6}\,\Omega \cdot \text{cm}$ for T = 300 K from [256] and the final resistivity for a 20 nm thick gold film on tantalum oxide are indicated on the y-axis.

the thicknesses for the oxide layers of 4.4 nm (TaOx) and of 4.8 nm and 7.2 nm (TiOx) of the samples used in this section.

4.6.2.2 Evaporation of gold top electrodes

The gold top electrode (6 mm· 9 mm) was prepared by thermal deposition at a rate of 0.3 nm/min under UHV conditions (base pressure $< 10^{-8}$ mbar) on top of the oxide layer. The bottom and the top electrode were contacted with micrometer thick silver strips similar to reference [255].

The twofold contacts of the top electrode allow an in situ monitoring of the gold film's resistance during its growth. In figure 4.38 the specific resistivity of the gold films (measured at 300 K) is plotted for thicknesses below 20 nm. The amount of gold evaporated on the sample was determined by a quartz balance. The x-axis in figure 4.38 is calibrated in the following way: three different amounts of gold were evaporated on a glass substrate. The gold films on the glass were scratched, and their thickness was determined by measuring the average scratch depth from AFM line scans performed across the scratches. Hence, the x-axis in figure 4.38 is normalized to the growth of gold on glass.

Obviously films grown on titanium oxide have a comparable high specific resistivity. Gold films on tantalum oxide reach the same resistivity values at thickness values which are 5.3 nm lower. At 20 nm thickness the specific resistivity of gold on tantalum oxide is $1.4 \cdot 10^{-5}\,\Omega \cdot \text{cm}$, this value is indicated on the y-axis of figure 4.38 as well as the bulk value for gold $\rho_{\text{bulk}}(\text{Au}) = 2.2 \cdot 10^{-6}\,\Omega \cdot \text{cm}$ [256]. The gold films on glass, which were used for the calibration of the x-axis in figure 4.38 show a much lower resistivity. Increasing the thickness from 6.5 to 8 nm leads to a considerable resistivity decrease of five orders of magnitude.

The apparent differences in the gold film growth on the three substrates can be partially discussed with classical arguments for the thermodynamics of interfaces. One crucial parameter determining the film growth is the surface free energy γ of the adlayer and the substrate.

Titanium oxide has a low value of surface free energy γ of $0.35\,\text{J}/\text{m}^2$ [257] whereas most metals show higher values. The values for metals range from $\gamma = 0.4\,\text{J}/\text{m}^2$ for lead [258] to $\gamma = 1.4\,\text{J}/\text{m}^2$ for gold [259, 260] and $\gamma \approx 2-3\,\text{J}/\text{m}^2$ for platinum [261, 262]. To the authors best knowledge, for tantalum oxide there is only one value in the literature of $\gamma = 0.28\,\text{J}/\text{m}^2$ which is cited in Ref. [257]. When the surface free energy of the oxide γ_{oxide}, the surface free energy of the metal γ_{metal} and the interfacial energy $\gamma_{\text{interface}}$ fulfill the inequality

$$\gamma_{\text{interface}} > (\gamma_{\text{oxide}} - \gamma_{\text{metal}}), \tag{4.22}$$

one expects cluster growth [263, 264]. With the above mentioned values the right hand side of equation 4.22 will be negative for gold on titanium and tantalum oxide. $\gamma_{\text{interface}}$ itself might only have a negative value if a chemical reaction happens between the gold and the oxide. This is unlikely for gold–titaniumoxide interfaces [265]. For gold–tantalum oxide interfaces the reaction of gold with the oxide is even more unlikely since the binding energy of tantalum oxide is higher. The binding energy for tantalum oxide is 201 kJ/mol [266], for titanium oxide the value is 47.7 KJ/mol [267]. Hence, for both oxide–metal systems $\gamma_{\text{interface}}$ can be expected to be positive leading to a tendency for cluster growth of gold on both oxides. This tendency should be weakest for gold on glass, since γ for glass surfaces is $\approx 1\,\text{J}/\text{m}^2$. This value is 2-3 times higher than the values for titanium oxide and tantalum oxide and leads to a lower negative value for the right hand side of equation 4.22. So, the comparatively low resistivity values for gold on glass can be motivated by the high γ value for glass surfaces.

All these considerations are valid only when one considers the surface free energies. But several complicated effects might influence this simple consideration.

- Adsorbates or defects on the oxide surface may influence the value of γ [268]. In our case a partial hydroxide coverage of the electrochemically built oxide cannot be ruled out [172, 173].

- A pure description in terms of interfacial energies might be insufficient since charge transfer processes between the adlayer and the substrate may induce a big interfacial stress [269, 270]. This stress (not included in equation 4.22) can influence the growth mode of the adlayer significantly [271].

In conclusion, I cannot debate on the differences between gold on tantalum oxide and titanium oxide with thermodynamic data (like values of surface and interface energies) from literature. A tentatively closed structure of gold on glass can be rationalized by the high γ value of glass. A more open structure of the gold film on titanium oxide with a higher fraction of voids is suggested. This argument will be supported by chemicurrent experiments in section 4.6.5.

4.6.3 Internal photoemission

4.6.3.1 Optical properties of the layer system

The optical properties of the layer system were calculated following the approaches of Edwards and Pepper [272, 273]. The Fresnel equations are evaluated for normal incidence, taking into account the bulk dielectric properties of the oxide layers [274–277] and of the metals [278]. Deviations from the bulk dielectric properties of the oxide and metal layers are not taken into account in the present work. For that purpose one would need an in-situ ellipsometric control of the films during growth [279, 280].

In figure 4.39 the absorptivities in each layer as well as the total transmissivity and reflectivity of the total system are shown. For low photon energies $h\nu < 2.5$ eV the absorptivity in the titanium respectively tantalum backelectrode is larger than in the gold top electrode. For both layer systems the absorptivity in the gold top electrode dominates for $h\nu > 2.7$ eV. The dotted vertical lines in figure 4.39 indicate the photon energy where absorptivities in both metal electrodes are equal. Tantalum oxide does not show any significant absorptivity over the whole energy range, whereas titanium oxide shows values of up to 10 % for 3.5 eV $< h\nu < 4.5$ eV. This is considered to be due to the lower band gap of titanium oxide of 3.6 eV [281]. Interband excitation processes will dominate in this energy range [282].

From a simple viewpoint, considering isotropic propagation of photoexcited carriers in the metal electrodes and over the oxide barrier, one should expect a polarity change of the photoinduced current at medium photon energies, e.g. $\approx$ $2.5 - 2.6$ eV, since the absorptivities in the top and in the back electrode are equal. Anisotropies in the propagation of the photoelectrons as for example induced by built-in internal electric fields may significantly alter the frequency dependence of the photon yield [283].

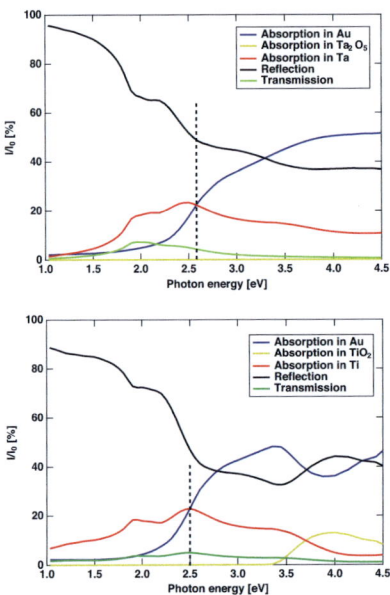

Figure 4.39: Absorptivities for the tantalum oxide (**upper graph**) and the titanium oxide (**lower graph**) based sensors. The dotted lines indicate the photon energies with equal absorptivities in the top and and in the back electrode.

4.6 Charge transport through thin amorphous TiOx and TaOx layers

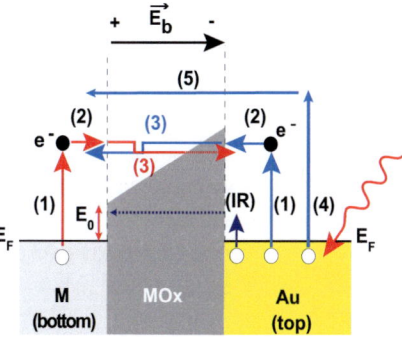

Figure 4.40: Schematic drawing of a metal–insulator–metal sensor and the built-in field $\vec{E}_b$. (IR) denotes photo excitation with $h \cdot \nu < 1\,\text{eV}$, the electron has to tunnel through the whole oxide. Process (1) denotes photo excitation for $1\,\text{eV} < h \cdot \nu < 2\,\text{eV}$. Process (2) assigns the transport to the interface. Process (3) denotes transport through the oxide with energy loss due to electron-phonon interaction leading to a lower tunnel probability for electrons going from the left to the right. Processes (4) and (5) denote over the barrier excitation and transport. E_0 denotes the lower height of the potential barrier at the backelectrode (titanium or tantalum)/oxide interface.

A layer system with a built-in field directed towards the top electrode is shown in figure 4.40. Scattering processes with optical phonons may occur when excited carriers pass the oxide layer. This leads to energy losses of several 100 meV [284, 285]. These energy losses hamper the transport of electrons travelling against the built-in field. After the scattering these electrons have a much lower tunnel probability, see path with red arrow from the left to the right hand side in figure 4.40. On the other hand, photoelectrons from the top electrode directly have to tunnel. Scattering processes during the subsequent transport in the oxide's conduction band then only have a minor effect because the transport to the conduction band of the bottom electrode is not hindered by a further barrier, see path with blue arrow from the right to the left hand side in figure 4.40.

It should be mentioned that electron-phonon (e-ph) scattering causes a higher energy loss in the oxide than in the metal since in the oxide the contribution of optical phonons dominates whereas acoustical phonons are the main scatterers in the metals. However, the above mentioned 'directional effect' of the oxides internal field will only work when the mean free path for the e-ph scattering in the oxide is in the range of or smaller than the oxide thickness. This seems to be valid since values of $0.4 - 0.6$ nm for the mean free path were found [286–288].

4.6.3.2 Optical set-up

As a light source for the high-energy photon studies I employed a Xe lamp in connection with a monochromator which produces a monochromatic beam of radiation with selectable photon energies in the range between 1.1 and 4.1 eV, i.e wavelengths of $300 - 1100$ nm. The linewidth of the beam was about 5 nm. The beam was guided to the MIM sensor by means of a 2 mm thick quartz fiber, whose end was mounted close to the sample to avoid that the illuminated area becomes larger than the active area as a result of beam divergence. The flux of photons impinging on the gold surface was calibrated by employing a silicon photodiode of known sensitivity.

For the low-energy photon studies (e.g. photon energies between 0.8 eV and 1.3 eV) the variable wavelength light source consisted of a quasi continuous wave (CW) high power white laser source, a Koheras Versa connected to a Czerny-Turner Monochromator (Triax180 by Horiba Jobin Yvon). Similar to the high-energy photon studies, a multimode fibre was used to deliver the monochromatic light to the MIM sensor. Again, care was taken to illuminate only the active area of the MIM sensor. Due to the low excitation energies, a sub-fA Source Measure Unit (Keithley Model K6430) had to be used to measure the subsequent small excited currents in the pA range at 0 V applied bias voltage. To avoid influence from external stray fields and external light, all measurements were carried out inside a shielding box.

Current measurements were performed by means of a HEKA potentiostat which incorporates a current-to-voltage converter with a conversion factor of 10 mV/pA. The Potentiostat did also supply the DC bias voltage applied between the top gold electrode and the bottom metal layer. The measured current is defined as the net number of detected electrons in the bottom electrode (Ta or Ti) per unit time. A negative current then means a net current of electrons flowing from the bottom into the top metal electrode. The photon-induced electron current is characterized by a yield

$$\gamma = \frac{\text{detected photo current}}{\text{impinging photon flux}}$$

defined as the average number of detected electrons induced by one incoming photon. The wavevector of the incident light was parallel to the surface normal of the samples.

4.6.3.3 Experimental results

The photoyield γ as a function of photon energy is shown in figure 4.41. It should be mentioned that for an applied bias voltage of 0 V the yield always shows positive values over the whole photon energy range. This means that a net photoelectron current flows from the top electrode to the back electrode. A current in the opposite direction, i.e. a polarity change cannot be observed, despite the fact that for energies $h \cdot \nu < 2.5$ eV more carriers are photoexcited in the bottom electrodes titanium or tanalum (see figure 4.39). Hence, the directional effect of the built-in electric field seems to work in both types of sensors at 0 V bias voltage.

In the upper graph of figure 4.41 the photoyield is plotted linearly. Strong increases of the photoyield are observed for $h \cdot \nu > 3.6$ eV for titanium oxide and $h \cdot \nu > 4$ eV for tantalum oxide. The inset in the upper graph shows the square root of the photoyield according to Fowler's analysis of photoemission [289]. This approach calculates γ as:

$$\gamma \propto (h \cdot \nu - \Phi)^2 \qquad (4.23)$$

where Φ is the workfunction in conventional photoemission or the height of the internal barrier in internal photoemission.

So, for high energies the square root of the photoyield γ is approximated by line

4 Measurements, results and discussion

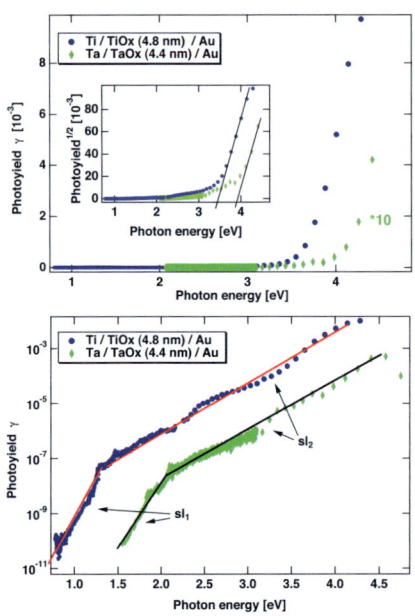

Figure 4.41: **Upper graph:** Photoyield for Ti–TiOx(4.8 nm)–Au and Ta–aOx(4.4 nm)–Au sensors plotted linearly. Inset shows the square root of the photoyield as function of photoenergy. The intersection of the line fits with the x-axis indicate the band gaps of the respective oxides. **Lower graph:** Logarithmic plot of the internal photoemission yield as a function of the incident photon energy for zero volt bias voltage. sl_1 and sl_2 denote the different slopes in the logarithmic plot.

fits, the intersection of the line fits with the energy axis is used as an estimation for the band gap as suggested in [290] and [281]. Values of 3.4 eV and 3.9 eV are found for titanium and for tantalum oxide.

The analysis of the data according to the laws of interfacial electron transfer in metal–liquid systems (sometimes cited as 5/2 law of photoelectrochemistry [291–294]) does not seem to be appropriate since the charge transfer ocurrs between two solid phases of our system.

In the lower graph the photoyield is scaled logarithmically. For both sensors regions with different slopes can be distinguished. For photon energies $h \cdot \nu <$ 0.8 eV no signal could be detected. In this case the carriers have to tunnel through the entire length of the oxide layer (see process (IR) in figure 4.40). From 0.8 eV to 1.3 eV a steep slope can be observed for titanium oxide samples, for tantalum oxide systems up to 2.0 eV. For these photon energies tunnel processes through the upper edge of a triangular tunnel barrier can be discussed similarly to the Fowler-Nordheim mechanism [295, 296]. For this purpose a parallelogram like barrier with

$$E_{\text{bar}}(z) = E_0 + \frac{z}{d_{\text{oxide}}} \cdot E_b \tag{4.24}$$

with different values of E_0 is assumed. E_0 is the barrier height at the titanium–titanium oxide or at the tantalum–tantalum oxide interface respectively, see figure 4.40. z denotes the coordinate normal to the system interfaces with $z = 0$ at the titanium–titanium oxide or at the tantalum–tantalum oxide interface. d_{oxide} denotes the thickness of the oxide. The barrier height at the opposed oxide interface (oxide–gold) is $E_0 + E_b$. E_b causes the unbalance of the oxides conduction band in figure 4.40. The unbalance of the conduction band evokes a built-in electric field $\vec{E}_b$ whose absolute value is given by

$$|\vec{E}_b| = \frac{E_b}{d_{\text{oxide}}} \tag{4.25}$$

The evaluation of the slope sl_1 can now be used to determine the unknown value for E_b by a simple iteration.

The tunnel probability $T(E)$ through the parallelogram like barrier $E_{\text{bar}}(z)$ can be expressed using the Wentzel-Kramers-Brillouin (WKB) approximation [297–299] as:

$$T(h \cdot \nu) = exp\left(-\kappa \cdot \Im\left(\int_0^{d_{\text{oxide}}} \sqrt{h \cdot \nu - E_{\text{bar}}(z)} dz\right)\right) \tag{4.26}$$

4 Measurements, results and discussion

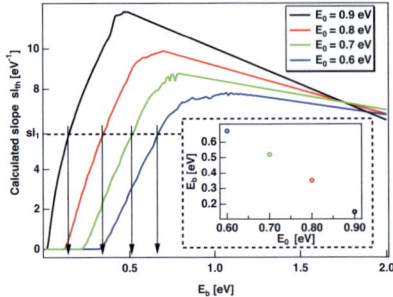

Figure 4.42: Calculated slope values according to equation 4.27. The experimental value $sl_1 = 5.8$ is denoted on the y-axis. Vertical arrows point to values of the built-in electric field, E_b, which fit to sl_1. Inset shows the combination of the barrier height at the metal–metal oxide interface, E_0, and E_b values which fit to experimentally determined sl_1.

with

$$\kappa = \sqrt{\frac{2 \cdot m}{\hbar^2}}$$

With equation 4.26 one can evaluate the partial derivative:

$$\left. \frac{\partial \log(T(h \cdot \nu))}{\partial h \cdot \nu} \right|_{h \cdot \nu_1}^{h \cdot \nu_2} = sl_{\text{th}} \qquad (4.27)$$

The theoretically found values sl_{th} have to be compared with the experimentally found slope values sl_1. Equation 4.27 is evaluated for $h \cdot \nu_1 = 0.9\,\text{eV}$ and $h \cdot \nu_2 = 1.3\,\text{eV}$ for titanium, while for tantalum $h \cdot \nu_1 = 1.6\,\text{eV}$ and $h \cdot \nu_2 = 1.9\,\text{eV}$ are used.

The calculated values of sl_{th} are shown in figure 4.42 for different values of E_0. The intersection of the calculated curves with the horizontal line given by $sl_{\text{th}} = sl_1$ gives possible values for the internal field E_b. The relation between the barrier height at the metal–metal oxide interface, E_0, and the built-in electric field, E_b, is shown in the inset of figure 4.42. The relation shows that the evaluation of the experimental results in figure 4.41 reached by the ratio of tunnel probabilities allows a determination of value pairs for E_0 and E_b. In table 4.3 the results from the evaluation of sl_1, E_0 and E_b are listed.

From the data presented here it is on the first glance difficult to discriminate between the value pairs E_0 and E_b fulfilling equation 4.27. However, for the tantalum–tantalum oxide–gold system one can find an argument for values of

4.6 Charge transport through thin amorphous TiOx and TaOx layers

	titanium	tantalum
sl_1	5.8	5.2
E_0 E_b	0.6 0.67	1.0 0.63
E_0 E_b	0.7 0.52	1.1 0.50
E_0 E_b	0.8 0.35	1.2 0.41
E_0 E_b	0.9 0.15	1.3 0.29

Table 4.3: Evaluation of the exponential dependence of the photoyield γ as function of $h \cdot \nu$ (see figures 4.41 and 4.42). The barrier height at the metal–metal oxide interface, E_0, and the built-in electric field, E_b, are value pairs [eV] fulfilling equation 4.27.

1.0 eV and 0.63 eV for E_0 and E_b respectively. The value of 0.63 eV is close to the difference of work functions for tantalum and gold of 0.7 eV, a value which was already found by R.H. Fowler [289]. We believe that these values coincide by chance, since the structure of the barrier is influenced for example by dipole layers at the interfaces and the hydroxide layer at the tantalum oxide–gold interface [172]. However, newer works about the dependence of the internal photoemission on the top electrode's thickness point also to a value of 0.6 until 0.7 eV for E_b [300]. So I think that a value of around 0.6 eV for E_b might be valid for both oxide systems in all likelihood.

For higher photon energies, i.e. $h \cdot \nu > 2\,\text{eV}$, γ shows again a linear behaviour in the logarithmic plot, but for both tantalum and titanium systems with a flatter slope. Interestingly the slopes in the two energy regions are very similar for both types of sensors. The flat slope ($sl_2 \approx 1.9 < 0.5 \cdot sl_1$) represents a weaker energy dependence. The evaluation procedure applied above following equations 4.27 and 4.26 leads to barrier heights of 9 eV. Obviously, this is not a meaningful value for a barrier height within our systems. The highest threshold energy is the band gap of the oxides of 3.4 and 3.9 eV as determined in the inset of the upper graph in figure 4.41.

An interpretation of sl_2 is at the present time not easy. Additionally the value of sl_2 is valid over a wide range of photon energies from 2 eV to 4 eV. Photoelectrons with these energies are transported in the conduction band. Thus, the low value of sl_2, expressing a weak energy dependence, might point to a scattering dominated transport in the conduction band. For tantalum oxide this transport does not depend on bias voltages up to 0.2 eV. Also both slope values sl_1 and sl_2 remain constant when the applied bias voltage is changed.

This is different for titanium oxide samples. Here the photoyield γ depends

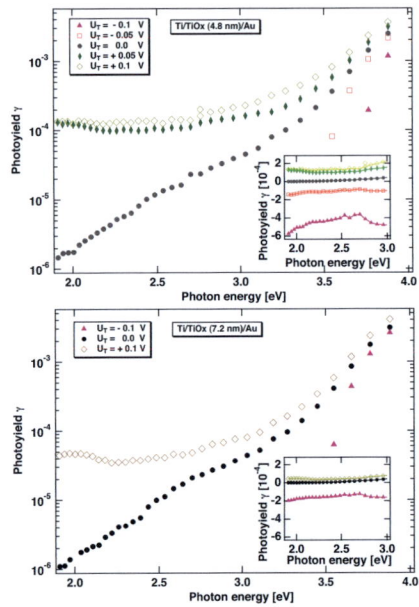

Figure 4.43: Logarithmic plot of the internal photoemission yield as a function of the incident photon energy for several bias voltages applied to the titanium back electrode. **Upper graph:** Ti–TiOx(4.8 nm)–Au sample. **Lower graph:** Ti–TiOx(7.2 nm)–Au sample. The inset shows a zoom of the photoemission yield plotted linearly for photon energies of 1.9 - 3 eV to clarify the polarity change of the photoinduced current, which can not be seen in the logarithmic plot.

clearly on the applied bias voltage, especially for the medium photon energy range from 2 eV to 4 eV as shown in figure 4.43. This figure shows the logarithmic plot of the photoyields γ of a Ti–TiOx(4.8 nm)–Au sample and a Ti–TiOx(7.2 nm)–Au sample. The photoyields are slightly smaller for the sample with the thicker oxide layer at 0 V bias voltage. The thickness dependence is much more pronounced for applied bias voltages. With a bias voltage of $+0.1$ V, γ values of 10^{-4} are found for the 4.8 nm sample, while for the 7.2 nm sample smaller values of $4 \cdot 10^{-5}$ are monitored. For a negative bias voltage even a polarity change of the photoinduced current can be observed. With photon energies $h \cdot \nu > 3.5$ eV, and negative bias voltage, the net photoelectron current flows from the gold top electrode to the titanium back electrode, the net current flows in the opposite direction for $h \cdot \nu < 3.5$ eV. The insets of figure 4.43 show linear plots of γ where negative values appear with nearly no dependence on the photon energy. With the negative bias voltage, γ shows also a pronounced thickness dependence for these photon energies. γ decreases from $-5 \cdot 10^{-4}$ for the 4.8 nm sample to $-2 \cdot 10^{-4}$ for the 7.2 nm sample.

Hence, for both bias voltages ± 0.1 V the exponential increase of the photoyield γ from $h \cdot \nu = 2\,\text{eV}$ to $h \cdot \nu = 3\,\text{eV}$ vanishes and is replaced by a more or less constant photoyield. The ratio of the photoyield $\gamma(U_T = 0.1\,\text{V})$ and $\gamma(U_T = 0.0\,\text{V})$ is shown in figure 4.44. The ratio seems to increase exponentially with decreasing photon energy down to 2 eV and reaches values of up to 80. An analysis with photoenergies below 2 eV was not possible. The photosignal induced by the chopped beam from the monochromator was overlaid by a significant fluctuation of the dark current signal. Experiments with a temperature stabilized set-up are under way to avoid these difficulties.

It should be mentioned that experiments in this range of photon energies (2 – 4 eV) cannot be explained with the previously discussed values for the barrier height E_0 or the internal field E_b. Carriers with these excess energies $h \cdot \nu > E_0 + E_b$ will be transported predominantly in the conduction band. Thus, it seems to be surprising that such a low bias voltage drastically influences the transport in the band. But one can think about a temperature activated hopping conduction. Activated conduction processes which can be influenced by small electric fields may happen in the band, similar to conduction on the edge of the conduction band as known in the so called Urbach tails [301–303]. The influence of the field is large for photon energies of 2 eV, i.e. conduction processes at the edge of the band. With higher photon energies the influence of the field becomes weaker and vanishes nearly for 4 eV (see figure 4.44).

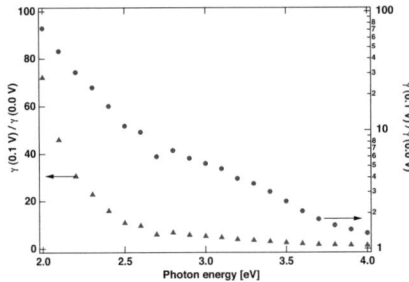

Figure 4.44: Influence of the bias voltage on the photoyield: Ratio of photoyields $\gamma(0.1\,\text{V})$ and $\gamma(0.0\,\text{V})$ for the Ti–TiOx(4.8 nm)–Au sample.

4.6.4 Bias-induced charge transport

4.6.4.1 Current driven by voltage ramps

Figure 4.45 depicts the typical current–voltage (I–U) characteristics of the Ta–TaOx–Au and Ti–TiOx–Au samples, recorded at a constant rate $dU/dt = 20\,\text{mV/s}$ in the voltage range between -0.2 V and +0.2 V. For an ideal metal–insulator–metal system, the tunneling current I_T should be very small for bias voltages U_T with $e \cdot U_T < E_0$, with E_0 being the lowest potential barrier in the heterosystem (see figure 4.40). In this case, the current response to the constant voltage ramp is only given by the time dependent (de-) charging current according to

$$C = \frac{Q}{U} = \frac{dQ/dt}{dU/dt} = \frac{I_{CH}}{dU/dt} \qquad (4.28)$$

Here, C is the capacitance introduced by the oxide layer, I_{CH} is determined as illustrated in figure 4.45. According to equation 4.28, the Ta–TaOx(4.4 nm)–Au, Ti–TiOx(4.8 nm)–Au and Ti–TiOx(7.2 nm)–Au samples have capacitances of 148 nF, 300 nF, and 235 nF, respectively. Taking into account the resistance of the gold top electrode $R_{Au} \approx 50\,\Omega$ and the titanium back electrode $R_{Ti} \approx 400\,\Omega$, one can derive a time constant τ of the system as $\tau = R \cdot C \approx 1\,\text{ms}$.

Ta–TaOx(4.4 nm)–Au and Ti–TiOx(7.2 nm)–Au exhibit, according to figure 4.45, a nearly ideal behavior. A strong deviation from this behavior is observed in the case of Ti–TiOx(4.8 nm)–Au. In this case an exponential dependence of the current on the bias voltage can already be seen. Since this exponential shape does not appear for the thicker titanium oxides, one could explain this by an increased tunnel probability through the thinner oxide. A detailed discussion will be given

4.6 Charge transport through thin amorphous TiOx and TaOx layers

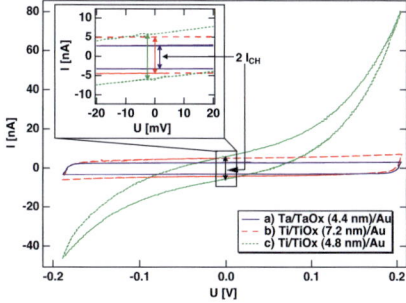

Figure 4.45: Current-voltage curves of a) Ta–TaOx(4.4 nm)–Au, b) Ti–TiOx(7.2 nm)–Au and c) Ti–TiOx(4.8 nm)–Au recorded at a constant rate $dU/dt = 20\,\text{mV/s}$. I_{CH} is the charging current according to equation 4.28.

in the section 4.6.4.3.

4.6.4.2 Transient shape of the sensor current

The transient shape of the sensor current after voltage steps allows the determination of the system's time constants in a more detailed way than the quasistationary current voltage method mentioned above. In the present section I use a system with a dynamically switching current-voltage converter. This converter is operated 0.1 and 1 s after the application of a voltage step. Thus, in one experiment one can measure the above determined time constants ($\tau \approx 1\,\text{ms}$) with currents of several mA as well as tunnel currents of some 100 pA for a long time. It is the aim of this section, to clearly separate the macroscopic time constants of the heterosystem as mentioned above (with a charge flow of some μC) from long term drifts (with a charge flow of some pC). The long term drifts may drastically influence the performance of chemicurrent and photocurrent measurements since the observed currents are in the pA range (see section 4.7 and reference [7,9].)

The experiment is conducted in the following way. The gold top electrode is always held on ground potential. The potentials indicated in the following figures are applied to the titanium backelectrode with the top fixed at ground potential. The initial potential E_{ini} is 0 V. The final potential U_{step} is positive on the titanium back electrode. We have chosen this polarization with an electron flow from the gold top electrode to the titanium back electrode since the net photoelectron current studied in section 4.6.3.3 flows in the same direction.

4 Measurements, results and discussion

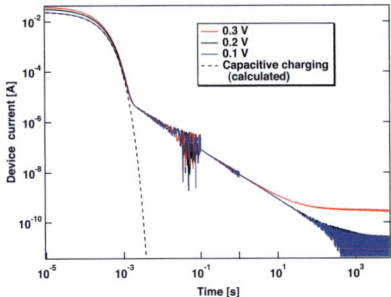

Figure 4.46: Transient shape of the sensor current after a voltage step from 0 V to a voltage U_{step} of 0.1, 0.2 and 0.3 V. Dotted line: calculated transient shape for $U_{step} = 0.1$ V according to equation 4.29. Device: Ti–TiOx(7.2 nm)–Au.

In figure 4.46 the transient shape of the sensor current after voltage steps from 0 V to the voltages U_{step} indicated in the figure are shown. The dotted line is a calculated charging curve of a capacitance ($C_{ox} = 250\,\mu F$) and a serial resistance ($520\,\Omega$) according to:

$$i(t) = \frac{U_{step}}{R_{serial}} \cdot e^{-\frac{t}{R_{serial} \cdot C_{ox}}}. \tag{4.29}$$

The values of C_{ox} and R_{serial} correspond to the capacity of the oxide layer and the sum of the resistivities of the titanium base and the gold top electrode. Obviously, the transient shape of the current is dominated by the capacitive charging in the first 10^{-3} s since the experimental curves and the calculated charging curve agree quite well. From $t = 10^{-3}$ s to $t = 10^1 - 10^2$ s a linear decay in the double logarithmic plot with a slope

$$\frac{\partial \log(I)}{\partial \log(t)} = -0.95 \pm 0.03 = \alpha$$

points to a power law like decay according to $I \propto t^\alpha$. Such a behavior is often found in thin vitreous oxide films [304–306] and was for the first time studied by J. Curie [307] and E.R. von Schweidler [308]. With $U_{step} = 0.3$ V a detectable current remains even after 10^3 s visible in our experiment.

With higher step voltages the remanent current increases and reaches a constant level after 10 s or 100 s for voltages of 0.5 V and 0.4 V respectively (see

4.6 Charge transport through thin amorphous TiOx and TaOx layers

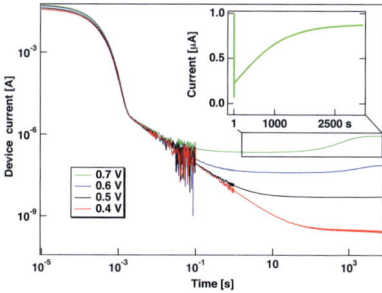

Figure 4.47: Transient shape of the sensor current after voltage steps $U_{step} = 0.4 - 0.7$ V. Inset: Linear plot of the current transient for $U_{step} = 0.7$ V from $t = 1$ s to $t = 3000$ s Device: Ti–TiOx(7.2 nm)–Au.

Figure 4.47). A different behaviour appears for $U_{step} > 0.5$ V. The current reaches a minimum value at around $t = 10$ s but then increases again, and settles a constant value after more than $5 \cdot 10^4$ s. The current transient for $U_{step} = 0.7$ V is plotted in the inset of figure 4.47 in linear scale. Its shape is proportional to $1 - e^{-\frac{t}{\tau}}$ where $\tau \approx 1000$ s. This second time constant τ of the system seems to be even longer for lower voltages (for example $U_{step} = 0.6$ V) since the current does not reach an asymptotic value even after $5 \cdot 10^4$ s. Experiments with $U_{step} > 0.7$ V lead to a remanent change of the sensor capacity, hence being disregarded in this work.

With negative values for U_{step} the shape of the current transients up to $t = 10$ s is partially similar. The charging domain up to $t = 10^{-3}$ s as well as the $\approx t^{-1}$ decay up to 10 s can be seen in the current transient. This decay is also followed by a smooth decrease to the asymptotic value for negative voltages. But, the subsequent increase of the current with large time constants does not appear for voltages -0.5 V $\leq |U_{step}| \leq -0.7$ V. Hence, a second time constant does not exist, when electrons flow from the titanium back electrode to the gold top electrode.

With tantalum sensors higher voltages up to 1.4 V in both polarities were applicable without changing their dielectric properties. After 10 s the currents remained comparably stable. A second time constant as observed for titanium did not show up. Hence, the current transients are not shown in the present work.

4 *Measurements, results and discussion*

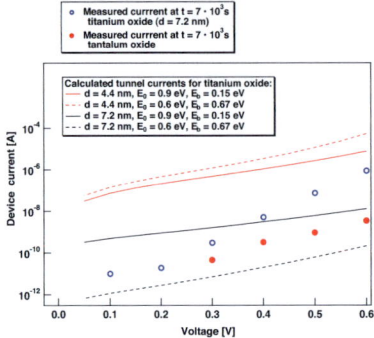

Figure 4.48: Steady state sensor currents as function of bias voltage for titanium (open blue circles) and tantalum oxide (filled red circles). Lines show calculated elastic tunnel currents with two different oxide thicknesses (each with two value pairs for the internal barrier height) for titanium oxide.

4.6.4.3 Steady state currents

The steady state currents observed for $t > 10^4$ s are plotted as a function of the bias voltage in figure 4.48 (open circles for titanium oxide, filled circles for tantalum oxide). The data are compared with elastic tunnel currents calculated for the titanium system for oxide thicknesses of 4.4 and 7.2 nm, as used in the experiment. As barrier parameters I used two value pairs from table 4.3, i.e. $E_F(\text{Ti}) - E_{CB}(\text{TiOx}) = E_0 = 0.6$ and $0.9\,\text{eV}$ and $E_F(\text{Au}) - E_{CB}(\text{TiOx}) = E_0 + |\vec{E}_b| \cdot d_{\text{oxide}} = 1.05$ and $1.27\,\text{eV}$, as suggested by the optical experiment. A band gap of 3.4 eV is assumed for titanium oxide, as evaluated in figure 4.41 and determined by photoelectron spectroscopy in the literature [281, 309, 310]. The tunnel probability through the barrier is evaluated using the WKB approach which is extended for a two-band system [311] in the present work. The current density impinging on the barrier is calculated by assuming two half spaces of free Bohr-Sommerfeld electron gases [14, 255, 312].

The magnitude of the measured sensor currents for the 7.2 nm thick titanium oxide samples coincides within 1.5 orders of magnitude with the calculated ones for small bias voltages. However, for $U_T > 0.3$ V the experimentally observed slope $\partial \log(I)/\partial U_T$ increases. For $U_T > 0.5$ V the measured values exceed the range of the calculated values for both barrier types. Hence, it is difficult to reconcile the experimental results with a constant barrier height. In contrast, a bias-induced additional lowering of the barrier in the course of the experiment must be considered for the explanation of the experiments. This bias-induced lowering might for example be driven by a charge exchange between the dipole layers at the two interfaces of the oxide. This process would of course bring a second time constant into the system. The assignment of two processes (fast dielectric charging and slow bias-induced altering of the barrier) is eased when one considers the transient behavior of the sensor current at higher bias voltages (see figure 4.47).

It should be mentioned that voltage step experiments were also performed with negative values for U_{step} (not shown in the previous section). In this case electrons flow from the titanium to the gold top electrode. The lower time constant could be also found. But the current always remained constant after 10 s. A second time constant with values $\tau > 10\,s$ did not appear.

The same is valid for tantalum based sensors. They only show a single time constant in the range of ms in both directions of current flow.

4.6.5 Chemicurrents

Photoexcitation is one possible way to produce carriers with an excess energy in metals. Carriers with an excess energy may also be produced in the direct vicinity of a metal surface by the stopping and neutralization of energetic atoms and ions [14, 15, 107, 108] or by non-adiabatic surface chemical reactions [6, 7]. We want to use the latter excitation channel to clarify the structure of gold on titanium oxide. One possible interpretation of the resistivity measurements in section 4.6.2.2 was a more open and heterogeneous structure of gold on titanium oxide compared to tantalum oxide. In previous experiments the dielectric properties of electronic sensors were found to be unchanged under exposure to a beam of atomic hydrogen. This was interpreted in terms of a completely covered oxide layer by the top metal gold [7]. The experiments were carried out with an intermitted atomic hydrogen beam under Ultra High Vacuum conditions with a tantalum oxide based sensor. The good repeatability and the unchanged dielectric properties of the sensor allowed to discuss the current answer of the tantalum oxide based sensors as a chemicurrent [7].

We repeated these chemicurrent experiments with Ti–TiOx–Au sensors. In figure 4.49 the transient shape of the induced sensor current can be seen, when the atomic hydrogen beam is switched on at $t = 10\,s$ and $t = 210\,s$ and switched off at $t = 50\,s$ and $t = 240\,s$. The repeated exposure between 210 s and 240 s clearly gives different results than the first exposure. Additionally a huge drift of the sensor current can be observed even when the atomic hydrogen beam is

4 Measurements, results and discussion

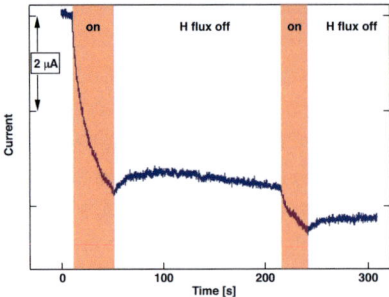

Figure 4.49: Transient shape of the induced sensor current of a Ti–TiOx(4.8 nm)–Au(20 nm) sensor. The atomic hydrogen beam is switched on in the red shaded time intervals. Experiments are accomplished at $T = 300$ K with an atomic hydrogen flux of $1 \cdot 10^{14}\,\mathrm{s}^{-1}$.

switched off. This behavior is attributed to an hydrogen induced change of the sensor. Since the gold thickness on tantalum and titanium oxide based sensors is the same, I attribute these apparent differences to a leakage of atomic hydrogen through the more open gold layer on the titanium oxide.

The ratio of the induced current of several μA and the incoming hydrogen atom flux would lead to yields of some 10 percent. However, a discussion of these sensor currents in terms of a chemicurrent is not appropriate from our point of view due to the observed instability of the sensor.

4.6.6 Conclusion

It is shown that internal photoemission is a versatile tool for the characterization of internal barriers in heterosystems when a wide range of photon energies is used. The existence of two different slopes (sl_1 and sl_2, see figure 4.50) in the logarithmic plot of the photoyield versus photon energy enables a clear discrimination between charge transport through a barrier, and above a barrier through the lower edge of a conduction band. The latter process appearing at higher photon energies shows a much weaker dependence on the photon energy than the transport through the barrier, meaning that $sl_2 < sl_1$ is fulfilled. The value of sl_1 allows the derivation of the barrier parameters by simple quantum mechanical considerations (equation 4.26 and figure 4.42).

A theoretical model for the transport in the lower edge of the conduction band

4.6 Charge transport through thin amorphous TiOx and TaOx layers

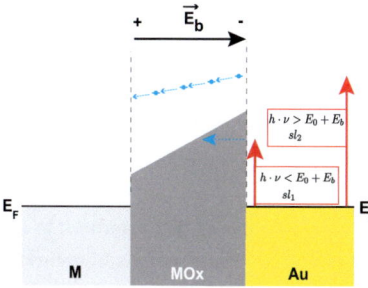

Figure 4.50: Model for transport of photoexcited carriers with energies between $h \cdot \nu > E_0 + E_b$ and $h \cdot \nu < E_0 + E_b$. sl_1 and sl_2 denote the slope as defined in figure 4.41. Transport through the barrier (sl_1) can be evaluated by quantum mechanical considerations. Transport through the lower edge of the conduction band (sl_2) might be a hindered transport with hopping events (symbolized by dashed arrows between weakly localized states).

(connected with sl_2) cannot be given in the moment. As a transport mechanism we suggest a hindered transport with hopping events between weakly localized states in the smeared out edge of the conduction band.

For titanium oxide samples this hindering can be cancelled out by a small bias voltage (< 0.2 V). This bias voltage leads to a sensitive sensor exhibiting a nearly broad band like response. This implies that $sl_2 \approx 0$.

For tantalum oxide based samples the dependence of the photoyield on the photon energy (sl_2) cannot be cancelled out by a bias voltage. This may point to a more durable carrier transport (lower influence by the bias voltage) in the lower edge of the conduction band in tantalum oxide based samples or to a lower density of weakly localized states in the lower edge of the conduction band.

The unusual behavior of the titanium oxide based samples can be found again in the experiments monitoring sensor currents after the application of voltage steps. The current transients showed complicated structures with two time constants for voltages $> 0.5\,V$. The lower time constant (some ms) is clearly associated with the equilibrium dielectric properties of the sensor ($\tau = R \cdot C$) and is bias-independent, whereas the second time constant (some 10 to 1000 seconds) depends on the bias voltage.

To conclude, titanium oxide based sensors offer a considerable higher photoyield compared to tantalum oxide based systems. So they seem to be promising candidates for the application as, for example, chemicurrent and photocurrent sensors at 0 V bias voltage. However, the complicated bias voltage dependence of the currents driven by internal photoemission or by bias voltages shows the limitations of the field of spectroscopy of internally excited electrons. For this purpose one needs barrier systems which depend in a simple way on the applied bias voltage [14, 15, 107].

For titanium oxide samples one has to pay attention to changes of the sensors during chemicurrent experiments. For reproducible chemicurrent experiments one will have to improve the wettability of the oxide for example by a surfactant. With a completely wetted oxide the high sensitivity of titanium oxide based sensors will allow the study of chemically induced electronic excitations with rather small energies with metal–insulator–metal sensors.

4.6.7 Appendix

In figure 4.51 AFM images of the 20 nm gold film of different sensors are shown. The roughnesses are $\approx$ 2 nm for the gold film on TaOx and SiOx. Whereas the gold film on TiOx shows some holes.

4.6 Charge transport through thin amorphous TiOx and TaOx layers

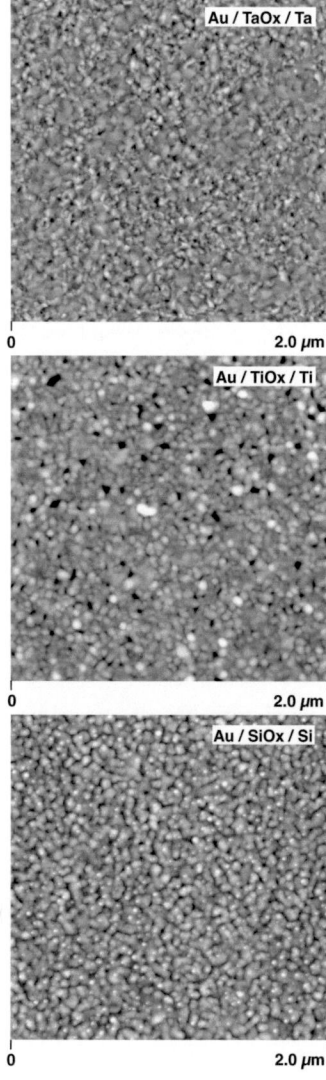

Figure 4.51: AFM images showing the gold film of different sensors. **Top graph** shows an Ta(20 nm)–TaOx(4 nm)–Au(20 nm) sensor, Au roughness ≈ 2 nm. **Middle graph** shows an Ti(20 nm)–TiOx(4 nm)–Au(20 nm) sensor, Au film with holes. **Bottom graph** shows an Si(500 μm)–SiOx(1 nm)–Au(20 nm) sensor, Au roughness ≈ 2 nm.

4.7 Photo-sensitive MIS sensors with stepped insulating layer[5]

Soon after the beginning of the studies on metal–semiconductor (MS) heterosystems, questions regarding the role of an interjacent oxide layer in the conduction properties of the sensor came up [313–319]. Provided that the oxide layer is characterized by a large optical band gap, the electrical conduction can be modeled by quantum mechanical tunneling processes [314, 315]. However, especially metal–insulator–semiconductor (MIS) sensors with an ultrathin oxide layer ($d < 5$ nm) are influenced by interface states at the semiconductor-oxide interface [320–324] and also midgap states, e.g. oxygen vacancies, which may dramatically change the electrical properties [325] of the sensors. Usually, in such sensors, the threshold energy (below which no photo-emission can be observed) is quite large, e.g. $3 - 4$ eV [290, 326–329], but it may be reduced by the interface states, mentioned before.

In this section, I present the experimental approach which was employed with the intention to rather improve the photo- and chemosensitivity [44, 330] of MIS sensors than to optimize their barrier properties. In particular, I studied the influence of the oxide thickness on the conduction properties of a Au-SiO$_x$-Si system by preparing an oxide layer of variable thickness on one wafer [331] with a localized electrochemical oxidation procedure (see figure 4.52).

As a measure of the photo-sensitivity I use the internal photoemission yield (or simply photoyield). It is defined as the net number of electrons detected in the Si electrode per number of photons incident on the Au surface. The individual sensors of the stepped MIS sensor were characterized by measuring the energy as well as the bias voltage dependence of the photoyield. Using a localized electrochemical oxidation technique, one can increase the photosensitivity of the silicon–silicon oxide–gold sensor with respect to the same sensor with 0 nm oxide thickness (silicon–gold sensor).

The four-step Au–SiOx–Si structure sketched in figure 4.52 was fabricated on a 2×1 cm^2 large piece from an n-type Si(111) wafer (7.5 Ω ·cm). The silicon surface was cleaned with isopropanol, hydrofluoric acid and MilliQ-water and oxidized in an electrolytic droplet cell (see section 4.3 and ref [85, 86]) using an ammonium acetate buffer electrolyte to minimize concurrent corrosion processes during the oxidation [87]. Different areas (typically of 0.04 cm^2) of the substrate were oxi-

---

[5]This section was published in Electrochem. Solid-State Lett. 12, H453 (2009) [81].

4.7 Photo-sensitive MIS sensors with stepped insulating layer

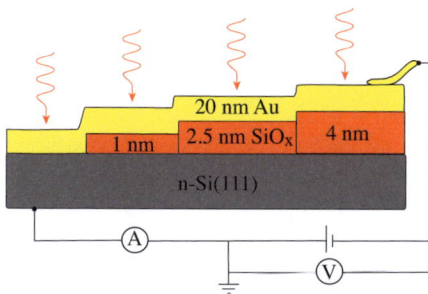

Figure 4.52: Schematic set-up of the Au–SiOx–nSi(111) sensor with a stepped oxide layer.

dized with the droplet cell. The desired oxide thicknesses were adjusted by varying the finite potentials (E_{final} = 2, 6 and 10 V) of the cyclovoltammograms (CVs). The different CVs were recorded with the initial potential fixed to -1.2 V and a scan rate of 0.1 V/s (Reference electrode Ag-AgCl). A stepped silicon oxide layer with terraces of about 1, 2.5 and 4 nm oxide thickness and equal area (0.6 cm^2) were prepared. Oxidation was performed under illumination of the droplet to provide a sufficient level of minority carrier generation which is necessary for the oxidation of n-doped silicon surfaces [332,333]. The value of 0 nm oxide thickness was reached by leaving the hydrofluoric acid treated surface unmodified. As a result, four different sensors were set-up on one sensor.

The CVs show at around -0.05 V ($E_{1/2}$ - half wave potential) a considerable current increase followed by a constant current of 150 μA/cm^2 for the oxidation of a hydrofluoric acid cleaned silicon surface (E_{final} = 2 or 6 V). When the electrode potential is stopped at E_{final}, the current drops exponentially to zero within a few seconds. The oxidation CVs itself can be used as a tool to check the homogeneity of the oxide. This is demonstrated by setting the droplet on the intersection of previously prepared 0 V, 1 V and 2 V thick oxide layers and applying a CV with E_{final} = 10 V. One can clearly see that all previously prepared thicknesses give a special shoulder in the cyclovoltammogram. Silanol formation [333] could not be observed in the CVs on n-type silicon.

The calibration of the oxide thickness was performed using ellipsometry and recording XPS sputter profiles. Film formation factors of 0.4 nm/V for the 6 V and 10 V oxide and 0.5 nm/V for the 2 V oxide were found (in agreement with ref. [121]). On top of the oxide, a gold film of 20 × 4 mm^2 was thermally evaporated

4 Measurements, results and discussion

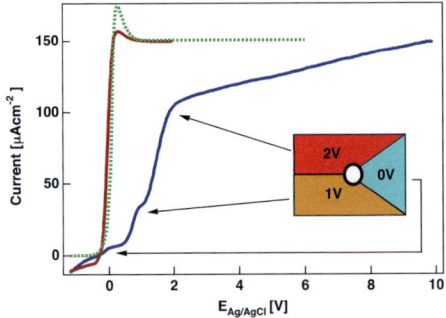

Figure 4.53: Cyclovoltammograms of silicon with the specified finite potentials on different areas of the sensor ($E_{initial} = -1.2$ V, $E_{final} = 2, 6$ and 10 V, $dE/dt = 0.1$ V/s). The cyclovoltammogram oxidized up to $E_{final} = 10$V was taken on the intersection of previously prepared 0 V, 1 V and 2 V oxide layers.

in high vacuum (base pressure $< 5 \times 10^{-8}$ mbar) at a low rate (0.3 nm/min) in order to ensure a homogeneous film thickness (see section 4.3). The gold film was contacted at the thicker oxide side to avoid short-circuits (see figure 4.52).

As a light source I employed a Xe lamp in connection with a monochromator which produces a monochromatic beam of radiation with a selectable wavelength in the range between 300 and 1100 nm (photon energies between 1.1 and 4.1 eV). The line width was about 5 nm. The flux of photons impinging on the gold surface was calibrated by employing a silicon photodiode of known sensitivity. The internal photoemission current was measured by means of a potentiostat.

In the following, the internal photoemission current is called 'positive' if a net electron current is flowing from the gold film to the silicon substrate. The same sign convention was used for the internal photoemission yield. The bias voltage is positive when the gold top electrode is positively biased with respect to the silicon back electrode. This is the conventional forward bias direction for an n-doped semiconductor–oxide–metal system.

In figure 4.54, the internal photoemission yield, induced in each of the four sensors of the MIS sensor, is plotted on a logarithmic scale as a function of the wavelength of the incident light. The most remarkable observation is that the "1 nm" MIS sensor not only exhibits the same broad-band photo-sensitivity as the "0 nm" (MS) sensor, but it is a factor of ten more photo-sensitive. So yields of $10^{-3} - 10^{-2}$ are measured on the "1 nm" sensor, but only of $10^{-4} - 10^{-3}$ on the

4.7 Photo-sensitive MIS sensors with stepped insulating layer

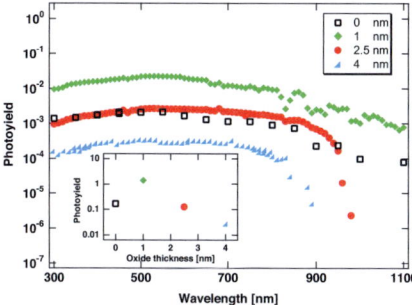

Figure 4.54: Logarithmic plot of the internal photoemission yield as a function of the wavelength of the incident light as determined for the four areas of the MIS sensor sketched in figure 4.52. The inset shows the photoemission yield at 635 nm as a function of the oxide thickness. All data taken at 0 V bias voltage.

"0 nm" sensor. Apparently, the oxide layer does not provide, as expected for an insulator, an additional potential barrier to the Schottky barrier present in the MS sensor, but leads, on the contrary, to a decrease of the effective barrier "felt" by the electrons excited in the Au layer. This behavior may be explained by the existence of interface states [334] and positively charged oxygen vacancies in the electrochemically formed oxide layer. In this case, the oxide rather exhibits the behavior of a p-type semiconductor, while the Si-SiOx structure may be viewed as an n-p junction. As a consequence, the transport of electrons from gold into silicon is facilitated by the presence of the built-in electric field in the p-n junctions.

With increasing oxide thickness, the smooth wavelength dependence is seen to persist only up to about 900 nm and 800 nm for the "2.5 nm" and the "4 nm" sensor, respectively. At greater wavelengths, the yield drops drastically, becoming negative at about 1000 nm and 900 nm, respectively. This indicates that, with increasing oxide thickness, a potential barrier for electrons develops in the oxide layer, which cannot be overcome by low-energy electrons (< 1.2 eV). As a result, the internal emission is dominated by holes (minority carriers) flowing from Au into Si or electrons flowing from Si to Au. The net yield decreases to $\approx -1 \cdot 10^{-4}$. In addition, the absolute yields measured on the "2.5 nm" sensor decrease by about two-, and on the "4 nm" sensor even by three orders of magnitude with respect to the "1 nm" sensor (inset of figure 4.54).

4 Measurements, results and discussion

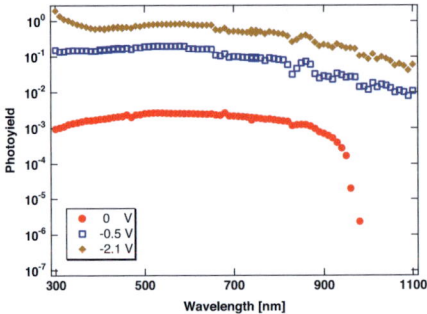

Figure 4.55: Internal photoemission yield as a function of wavelength, plotted for selected negative bias voltages, as measured on the "2.5 nm" step of the MIS sensor sketched in figure 4.52.

The extra barrier introduced by the thicker oxides can be, however, counterbalanced by applying a negative bias voltage to the Au electrode. As shown in figure 4.55, already -0.5 eV are sufficient to extend the broad-band sensitivity region of the "2.5 nm" sensor up to (at least) 1100 nm. In addition, negative biases increase the absolute value of the yields to $10^{-2} - 10^0$.

Comparative conductivity-voltage characterization of the sensors show small values for negative bias voltages and a strong increase for positive bias voltages similar to previous studies [315, 320]. This means that the increase of the photoyield for negative bias voltages is clearly observed in reverse voltage direction for all sensors.

In conclusion I presented an experimental approach based on localized electrochemical oxidation for the fabrication of MIS sensors. This procedure allows the production of several MIS sensors of different oxide thicknesses on a single silicon substrate, enabling in future also the production of sensor arrays. The different sensors corresponding to each step of the underlying oxide layer were investigated separately by internal photoemission. The MIS sensors show a maximum (at around 1 nm oxide thickness) in the photoemission yield when scaling the oxide thickness from 0 to 4 nm, similar to [321]. Experiments with a greater variety of E_{final} values of the CVs will allow further optimization of the sensors photosensitivities. The broad-band sensitivity is probably due to a broad distribution of states across the gap as discussed in [334]. Results of a detailed investigation regarding the energy levels of such states using equally prepared sensors with uniform oxide thickness, by means of capacitance-voltage and current-voltage measurements, will be reported in a later section (4.8).

4.8 Transport of excited holes through MIS sensors[6]

4.8.1 Introduction

Weak electronic excitations may occur on metal surfaces by chemical reactions. With some exceptions in the case of strong exothermal reactions [335] one cannot detect these excitations as an electron current leaving the reacting surface into a vacuum chamber. The electronic work function must be overcome by the excited electrons. Hence, only distribution of electrons with a significant density of states $\rho(E)$ above the electronic work function Φ can be observed from the detection of an external electron current.

For the detection of weak electronic distributions with $\rho(E) < \Phi$ thin film electronic sensors with an internal barrier $\varphi < \Phi$ were introduced in the last decade. The sensors are either metal-semiconductor [6], metal–insulator–metal [7] or metal-insulator-semiconductor [44, 336–338] sensors.

In the present section I focus on the transport of excited electrons in electrochemically prepared **m**etal-**i**nsulator-**s**emiconductor sensors (MIS). For electronic excitation we use either monochromatic light or a low energy Ar$^+$ ion beam. The optical excitation in multilayer systems often comprises more than one layer, when the thickness of the layers is comparable to the optical penetration length [14]. Thus, for comparison Ar$^+$ ion induced excited carriers are studied. With a kinetic energy $E_{kin} = 200$ eV, as used in this study, the penetration depth is only 2 nm [339]. Hence the excited carriers are produced in the first ten atomic layers of the metal top electrode.

The spatial separation of possible excitation processes is illustrated in figure 4.56. Due to the small thickness of the metal films ($d = 20$ nm Au or $d = 7$ nm Pt) photo induced electron hole pairs may be excited in the space charge layer of the semiconductor (its thickness denoted as d_{SCL}) for photon energies $h \cdot \nu > E_{gap}$ (process 1 in figure 4.56). d_{SCL} can be estimated [47] by

$$d_{SCL} = \sqrt{\frac{2 \cdot V_s \cdot \epsilon_{Si} \cdot \epsilon_0}{e \cdot N_d}}. \quad (4.30)$$

Equation 4.30 gives $d_{SCL} = 1.5\,\mu m$ when $\epsilon_{Si} = 12$, $V_S = 1V$ and $N_d = 6 \cdot 10^{14}$ cm^{-3} for a 7.5 $\Omega \cdot$ cm n-type silicon.

Electrons and holes excited in the silicon are separated by the electric field in the space charge layer. The electrons flow back into the bulk of the semiconductor

---
[6]This section is slightly changed from the version to be submitted to Phys. Rev. B.

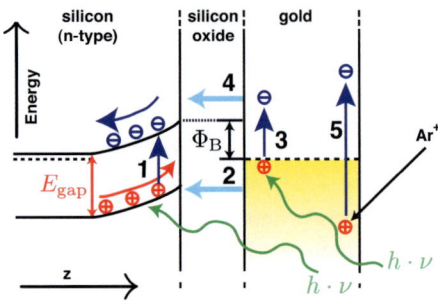

Figure 4.56: Excitation processes in MIS systems with n-type semiconductor induced by light or Ar$^+$ ions. 1: Photoexcitation in the space charge layer of the semiconductor with $h \cdot \nu > E_{gap}$. 2: Subsequent neutralization of the excited hole by a metal electron. 3: Photoexcitation in the metal with $h \cdot \nu > \Phi_B$ 4: Subsequent transport of the excited metal electron to the semiconductor. 5: Electronic excitation of electrons and holes in the topmost region of the top metal film by an Ar$^+$ ion beam. (z-axis is not in true scale, thickness of the space charge layer d_{SCL} estimated using equation 4.30.)

and do not contribute to the current measured in the sensor. The holes will be collected at the semiconductor oxide interface and then can be neutralized by electrons tunneling from the top metal electrode to the silicon substrate (process 2 in figure 4.56) and thus constitute a sensor current (electrons from top to bottom).

For smaller photon energies excitation processes in the silicon are excluded due to the indirect band gap of 1.1 eV. For energies with $\Phi_B < h \cdot \nu < E_{gap}$ excited electrons released in the top metal film (process 3) may overcome the Schottky barrier Φ_B when the transmission of the oxide is sufficiently large (process 4). This would also lead to a net electron current from the metal to the silicon layer (from the top to the bottom).

An excitation solely in the first atomic layers can be obtained for example by low energy Ar-ions. This is denoted as process 5 in figure 4.56. The distribution of excited electrons and holes produced by this low energy ions can be well described by a Boltzmann distribution with electronic temperatures of a few 1000 K [14,15]. Two different processes may evoke an ion induced sensor current:

1. The excited holes can travel to the oxide-metal interface and can be neutralized there by electrons from the valence band of the silicon, provided that the oxide has a sufficient transmission probability in this energy range.

4.8 Transport of excited holes through MIS sensors

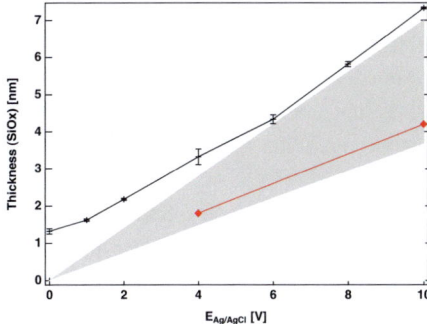

Figure 4.57: Silicon oxide thickness derived by a) ellipsometry using silicon oxide bulk optical data (black line) b) oxide formation factors [121] (gray shaded area) and c) XRD-sputter profiles (red line).

This would be possible, when midgap states are present in the vicinity of the valence band of the oxide.

2. The excited electrons may come to the oxide-metal interface and can be detected as a current with the sensor. For this process midgap states in the vicinity of the oxide's conduction band would be necessary.

In this section I study the dependence of the photon and ion induced carrier transport on the bias voltage of the sensor, the thickness of the oxide and the energy of the photons and ions. Thereby we derive the pathways and processes used by excited carriers in MIS sensors.

4.8.2 Experimental

The procedure of the electrochemical preparation of the silicon wafer and the oxide film with stepped oxide thickness was previously described in section 4.7 [81]. The band gap of the chemically prepared oxide films with a value of $9 \pm 0.5\,\text{eV}$ was investigated using electron energy loss spectroscopy [340]. The characterization of the oxide films especially with respect to the evaluation of the film thickness is performed with two independent methods, ellipsometry and XRD-sputter profiles.

XRD-sputter profiles were obtained for two silicon samples, oxidized with electrode potentials of 4 and 10 V. Thicknesses values of 1.8 and 4.2 nm were found (red line in figure 4.57). Experiments with lower oxidation potentials were not attempted since the sputter processes do not only release oxygen into the vacuum, but also drive oxygen into the deeper layers of the silicon. Hence, the method is unsuited for our experiments with very thin oxide films. Linear extrapolation of the results for 4 and 10 V, suggests a thickness of 1 nm for 2 V electrode potential. The slope of the linear extrapolation has a value of 0.43 ± 0.05 nm/V which is named film formation factor. Literature values of the film formation factor vary between 0.4 nm/V and 0.7 nm/V [121]. The conceivable thickness values with these film formation factors lie in the gray shaded area. Experimental results obtained via ellipsometry, using the optical bulk values of silicon dioxide (black squares in figure 4.57), can be found above the upper edge of the gray shaded area given by literature values. Deviations become even larger for a thickness smaller than 2 nm. This is not surprising since the model uses the bulk values assuming a Si^{4+} oxidation state. For thinner oxides one has to keep in mind that a mixture of different oxidation states exists even if thin oxide films are prepared under high vacuum conditions [341]. This will lead to the increasing deviation of the optical constants from the bulk values, delivering wrong values in the analysis of the oxide thickness. Hence, the oxide thickness values given by the XRD-sputter experiments, which are consistent with the literature values for film formation factors, are used in this section.

After the electrochemical preparation the oxide is covered with a 20 nm thick gold film, 20 nm thick gold pads, or a 7 nm thick platinum film, all deposited in a UHV chamber.

For the current voltage measurements on the MIS samples a sub-fA source measurement unit (SMU, Keithley Model K6430) was used. The time behavior of the corresponding voltage sweep was controlled via a Labview program. To avoid influence from external stray fields and external light, the measurements were carried out inside a shielding box. The electronic contacts of the sensors where established by pressing a gold wire with a small droplet of indium on an already established metal top contact. The bias voltage is counted positive when the Au top electrode is positively biased with respect to the Si back electrode. This is the conventional forward bias direction for an n-doped semiconductor–oxide–metal system. Current voltage dependences were taken dynamically with different scan rates.

For the capacitance voltage measurements on the MIS samples a LCR meter (HP Model 4284A) was used. A needle prober inside a dark box was used for establishing the electrical contacts. For monitoring the bias voltage dependence of the capacitance we waited a sufficient time to allow the equilibration of the system.

The same light source and potentiostat set-up as in section 4.7 was used. The potentiostat did also supply the DC bias applied between top gold electrode and the bottom metal layer. As in section 4.7 the internal photoemission current is called 'positive' if a net electron current is flowing from the gold film to the silicon substrate. The presented photoemission current is a net current, taking into account the difference of the measured signals between illuminated and dark experiment. The same sign convention holds for the internal photoemission yield. A negative photoyield means a net electron current from the silicon backelectrode to the metal top electrode or a net hole current vice versa.

4.8.3 Electronic sensor properties

4.8.3.1 Current voltage plots

In figure 4.58 current vs. voltage plots recorded for the MIS sensors with 1 and 4 nm oxide thickness and for three different scan rates are shown. For both sensors the observed current is much larger for positive bias voltages. This can be expected since this is the normal forward current direction for an n-type semiconductor. The backward and forward current traces show a hysteresis which is caused by charging the capacitance of the sensor. The hysteresis between the forward and the backward current is a charging current I_{CH} which can be used according to

$$C = \frac{dQ}{dU} = \frac{dQ/dt}{dU/dt} = \frac{I_{CH}}{dU/dt} \qquad (4.31)$$

for the determination of the capacitance C. Values of $\approx 25\,\text{nF}/\,\text{mm}^2$ for the 1 nm thick oxide and $\approx 18\,\text{nF}/\,\text{mm}^2$ for the 4 nm thick oxide samples are obtained when the hysteresis is evaluated at 0 V. A comparison with a simple capacitor model is not appropriate since the sensor capacitance is built by the oxide layer and the space charge layer in the semiconductor. The latter one depends clearly on the bias voltage. Additionally the silicon oxide itself consists of several different layers as mentioned above. Thus, simple equivalent circuits might not be sufficient to describe the sensors. A detailed capacitance study by an impedance meter using an alternating voltage at variable bias voltages is presented later in section 4.8.3.2.

Additionally the current-voltage plots show comparable current densities for both oxide thicknesses. The value of $60\,\text{nA}\,\text{mm}^{-2}$ in forward direction at 0.2 V bias voltage applied on the gold electrode for the 4 nm oxide sample is compa-

4 Measurements, results and discussion

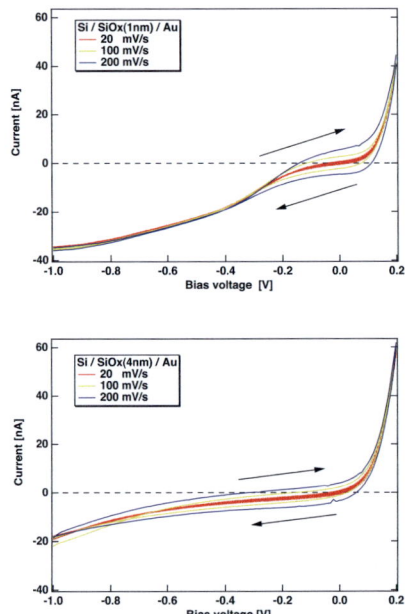

Figure 4.58: **Upper graph:** Current-voltage plots of the MIS sensors for 1 nm oxide thickness. **Lower graph:** Current-voltage plots for 4 nm oxide thickness. Both plots are shown for three different scan rates. The area of gold electrode is 1 mm^2 (see inset in figure 4.60).

rable with values in the literature for 3.5 nm thick oxides (2 nA mm^{-2} at 0.2 V in forward direction [342]). Since the current density is more or less the same for the 1 nm oxide sample, one can state that a pure tunnel process through the oxide barrier cannot be the limiting factor for the current in forward direction. The derivation of a barrier height by the current-voltage curves is therefore not possible. Further information about the barrier height will be gained by the photoemission experiments shown in section 4.8.5.2.

The same seems to be valid for the current densities in the backward direction. But the 1 nm oxide sample shows a diminishing shift between the current in forward and backward scan direction of the bias voltage applied on the gold electrode (between -0.2 and -1.0 V). This would point to a decrease of the capacitance for these bias voltages on the gold electrode, as is typical for n-type sensors [314, 315]. The voltage dependence of the capacitance will be studied in more detail in the follwing section 4.8.3.2.

4.8.3.2 Capacitance voltage plots at different frequencies

Figure 4.59 shows the capacitance–voltage characteristics of Si–SiOx–Au sensors with 1 and 4 nm oxide thickness for three probe frequencies. It should be mentioned that the forward and backward scan of the bias voltage applied on the gold electrode is shown in contrast to older studies in the literature [315]. This enables later a clear assignment of the oxide charge [313, 343]. A clear dependence on the bias voltage as well as on the frequency can be seen. The capacitances show the characteristic increase from the depletion region toward 0 V bias voltage and a saturation in the accumulation regime C_{max} with values from $C_{max,3} = 6 \cdot 10^{-10}$ F to $C_{max,1} = 1.3 \cdot 10^{-8}$ F. In the depletion regime the sensor capacitance shows a value of $C_{min} = 2 \cdot 10^{-10}$ F and is nearly independent on the frequency. The ratio $C_{max,i}/C_{min}$ spans a range from 3 (1 MHz) to 60 (20 Hz) with decreasing frequency due to a lowering of C_{max} with increasing frequency.

The frequency dependence of C_{max} is known to result from the relaxation times τ of the charge transfer between the semiconductor in accumulation and the oxide layer [334]. This charge transfer comprising interface states induces capacitances which depend on the applied bias voltage (denoted in the inset as C_{IS} of the lower graph of figure 4.59). R_{IS} denotes the serial contact resistance. The charge transfer to the interface states depends on the frequency and can be treated in a classical relaxation approach [344] leading to the following expression

$$C_{IS}(\omega) = \frac{C_{IS}}{1 + \omega^2 \cdot \tau^2} ,$$

4 Measurements, results and discussion

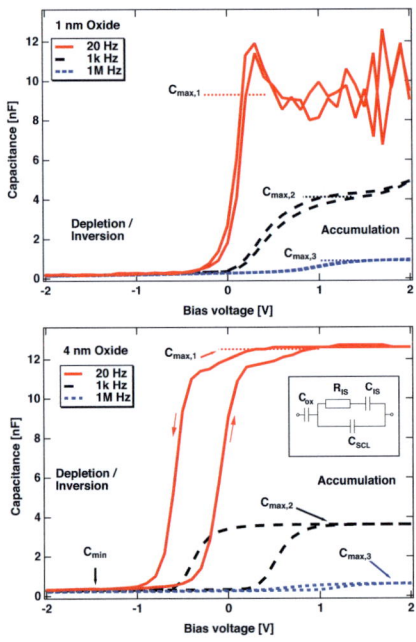

Figure 4.59: Static capacitance–voltage plot measured using the 1 nm thick (**upper graph**) and the 4 nm thick (**lower graph**) oxide area of the Si–SiOx–Au sensor for 20 Hz, 1kHz and 1 MHz. Inset shows equivalent circuit representing the capacitance of the oxide C_{ox}, interface states C_{IS} and the depletion layer C_{SCL}. Sample set-up as in figure 4.60. Direct of forward and backward scan indicated by red arrows.

4.8 Transport of excited holes through MIS sensors

where τ is the above mentioned relaxation time.

Charge transfer in the semiconductor occurs only in the space charge layer, the capacitance evoked by this process is C_{SCL}. The two semiconductor induced capacitances can be regarded as connected in parallel:

$$C_{SCL} + \frac{C_{IS}}{1+\omega^2 \cdot \tau^2} . \qquad (4.32)$$

The capacitance of the whole equivalent circuit is then the serial connection of C_{ox} and the capacity represented by the terms in equation 4.32.

Using this simple model with $C_{IS} = 2.2 \cdot 10^{-8}$ F, $C_{ox} = 3 \cdot 10^{-8}$ F, $C_{SCL} = 2 \cdot 10^{-10}$ F, $\tau = 300\,\mu s$ for the 4 nm oxide system we get calculated values which are compared with the experimental values in table 4.4. The values for C_{IS}, C_{ox} were chosen by fulfilling $1/C_{max,1} = 1/C_{IS}(\omega) + 1/C_{ox}$ for $\omega = 2\pi \cdot 20\,Hz$. C_{SCL} was set to be $C_{SCL} = C_{min}$. The value of $\tau = 300\,\mu s$ was chosen to fit the ratio $C_{max,1}/C_{max,2}$. The difference between the experimentally determined values of 12.6 nF, 3.6 nF and the calculated values 12.6 nF, 4.0 nF is smaller than 10 %. The discrepancy between calculated and experimental values is larger for $f = 1\,MHz$ where nearly a factor of 3 lies between experimental and calculated values. This means that the assumption $C_{SCL} = C_{min}$ is a too rough approach. In accordance with literature C_{min} can be underestimated by a factor of 3, when the probe frequency is a factor of 10 too low [345]. However, an impedance study with 10 MHz probe frequency leads to unreliable results due to inductivity problems with our present Ultra High Vacuum set-up.

For the samples with 1 nm thick oxide we took for the calculation simply a factor of 4 higher value for C_{ox} (i.e. 120 nF), the same time constant τ and the same value for C_{IS}. By that one gets the results shown in table 4.4. Even with these simple assumptions one gets a good agreement for $f = 1\,kHz$ between the experimental value 4.3 nF and the calculated value 4.5 nF. For 20 Hz the discrepancy is much larger (factor 2) than for the sample with the 4 nm thick oxide. Similar deviations appprear for 1 MHz between experimental and calculated values, a factor of 4.5 in this case. Both discrepancies might be due to the fact that the separation between C_{IS} and C_{ox} as introduced in the equivalent circuit is no longer meaningful for the sample with the 1 nm thick oxide.

Since the intermediate frequency case ($f = 1\,kHz$) is best described by our simple model, we chose this case to derive the interface state density by the capacitance voltage plots.

Table 4.4: Data obtained via calculation (see explanations in text) and experiment (compare figure 4.59).

oxide thickness [nm]	frequency	calculated values [nF]	experimental values [nF]
4	20 Hz	12.6	12.6
4	1 kHz	4.0	3.6
4	1 MHz	0.2	0.62
1	20 Hz	18.4	9.8
1	1 kHz	4.5	4.3
1	1 MHz	0.2	0.9

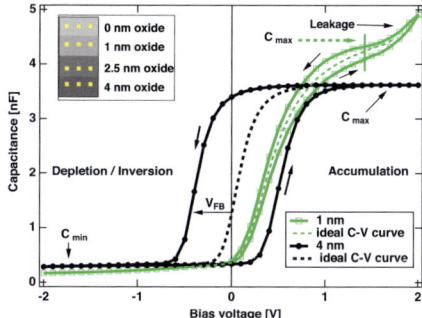

Figure 4.60: Capacitance–voltage behavior of the 1 and 4 nm thick oxide area of the Si–SiOx–Au sensor, measured with 1 kHz. Gold top electrodes are 1 ∗ 1 mm gold pads on the differently oxidized areas of the silicon piece as shown in the inset. Dashed curves (calculated by an algorithm shown in the appendix) symbolize the ideal CV-curves without any oxide charge. C_{max} indicates the capacitance value in accumulation. V_{FB} denotes the flat band voltage for the 4 nm thick oxide. Inset shows the sample set-up with four differently oxidized areas.

4.8.4 Capacitance voltage plots at different thicknesses

4.8.4.1 4 nm thick oxides

Figure 4.60 depicts the capacitance–voltage behavior of the Si–SiOx–Au sensor for two different oxide thicknesses measured at 1 kHz. The sample with the 4 nm thick oxide shows a typical behavior for a MOS capacitance. In the depletion region the capacitance is small, whereas it is big in the accumulation region. For bias voltages larger than 1 V a saturation value of the capacitance C_{max} can be observed. This saturation value of the capacitance is determined determined by the oxide layer [346]. Via

$$\epsilon_{rel.} = C_{max} \cdot d / (\epsilon_0 \cdot A),$$

$C_{max} = 3.6$ nF/mm^2 (see figure 4.60), and 4 nm for the oxide thickness, one obtains a value for the dielectric permittivity $\epsilon_{rel.}$ of 1.6. In the depletion region one can evaluate $C_{min} = 0.32$ nF/mm^2. By

$$\frac{1}{C_{SCL}} = \frac{1}{C_{min}} - \frac{1}{C_{max}} \qquad (4.33)$$

one arrives at $C_{SCL} = 0.35\,\text{nF/mm}^2$.

The thickness of the space charge layer in depletion can be determined by

$$d_{SCL} = \epsilon_0 \cdot \epsilon_{Si} \cdot A / C_{SCL}$$

to $3.3 \cdot 10^{-7}$ m. $\epsilon_{Si} = 12$ was assumed hereby. This value agrees quite well with literature values which quote $4 \cdot 10^{-7}$ m for a dopant concentration of $N_d = 10^{16}\,\text{cm}^{-3}$ [347].

d_{SCL} is larger than the Debye radius L_D in the semiconductor which is given by

$$L_D = \sqrt{\frac{\epsilon_0 \cdot \epsilon_{Si} \cdot k_B \cdot T}{e^2 \cdot N_d}} \qquad (4.34)$$

and can be evaluated as $L_D = 1.7 \cdot 10^{-7}$ m with $T = 300\,\text{K}$ and the above mentioned parameters.

The CV-curve shows a hysteresis which corresponds to an oxide charge Q_{ox} which can be determined by

$$Q_{ox} = \int C_1(V)\,dV - \int C_2(V)\,dV \qquad (4.35)$$

The dashed curve in figure 4.60 shows an ideal CV-curve without any oxide charges. The shape of this curve is derived by a mathematical algorithm presented in the appendix. This allows us to evaluate the oxide charges $+Q_0$ and $-Q_0$ without taken the uncharged oxide ($C_{ideal}(V)$) into account (compare reference [348]). The value of $+Q_0 = 1.5 \cdot 10^{-9}\,\text{C/mm}^2$ is determined by integrating the capacitance on the left hand side of the ideal CV-curve as shown in the appendix and in [348]. $-Q_0$ can be evaluated by integrating the right hand side of $C_{ideal}(V)$.

The flat band voltage can be evaluated by the intersection of the ideal CV-curve with the ordinate. The flat band voltage corresponds to the shift in bias voltage, as shown as horizontal arrow and its length denoted as V_{FB} in figure 4.60. This value of 0.45 V is the flat band voltage for the 4 nm oxide system. By solving

$$V_{FB} - \Delta\Phi = -\frac{Q_0}{C_{max}} \qquad (4.36)$$

one can derive the work function difference $\Delta\Phi$ [349] between the silicon and the gold electrodes to 0.05 eV [348]. Such a low value of $\Delta\Phi$ is confirmed by literature values. The literature value of $\Delta\Phi = 0.23\,\text{eV}$ can be obtained when one takes the work function value for polycrystalline gold of $\Phi_{Au} = 4.83\,\text{eV}$ [350]

and for silicon of $\Phi_{Si} = 4.6\,\text{eV}$ [351, 352]. The discrepancy of 0.18 V between the experimentally derived work function difference $\Delta\Phi = 0.05\,\text{eV}$ and the literature value of $\Delta\Phi = 0.23\,\text{eV}$ might be due to a lowered work function of our thin gold films.

4.8.4.2 1 nm thick oxides

The 1 nm thick oxide sample shows overall a slightly higher capacitance. Additionally the CV-curve is superimposed by a capacitance increase at bias voltages $> 1.8\,\text{V}$. This increase is attributed to a leakage induced polarization caused by the high field strength of 1.8 GV/m. Similar phenomena are found also for MIS sensors with 5 nm interstitial thermal oxide [353]. Between the regime of polarization induced capacitance increase and leakage induced capacitance increase a transition region at 1.5 V bias voltage can be found. This region can be characterized by $\partial^2 C/\partial V^2 = 0$ (turning point in the C-V plot) and is indicated in figure 4.60 by the horizontal green arrow. With a capacitance C_{max} of 4.35 nF/mm^2 and an oxide thickness of 1 nm one gets a value of 0.5 for $\epsilon_{rel.}$. In this case one has to be aware of the possible experimental errors in determining the oxide thickness as shown in figure 4.57. If the oxide thickness of 1 nm is replaced by a value of 2 nm, one would get $\epsilon_{rel.} \approx 1$. Also this value is far smaller than the ones, which are typically found for thin silicon oxide films by theoretical calculations [354–356]. The positive and negative sweep show a very small hysteresis. This small hysteresis points to a small oxide charge Q_{ox}. The charge density Q_0 is 0.2 nC/mm^2. For the 1 nm oxide a flat band voltage V_{FB} of nearly zero Volt is evaluated in accordance with literature data [315]. The low value of $\Delta\Phi$ found for the 4 nm thick oxide samples is found for the 1 nm thick samples as well. With the values of Q_0, C_{max} and using equation 4.36 one gets a value of $\Delta\Phi = 0.05\,\text{eV}$.

The Q_0 values of the CV-curves can be used to derive an areal density and a volumic density of states. These values, the flat band voltages V_{FB} and the values for $\Delta\Phi$ are shown in table 4.5. Values of around $10^{12}\,\text{cm}^{-2}$ for the areal density are in good agreement with literature values of $10^{11} - 10^{12}\,\text{cm}^{-2}$ [246, 357, 358].

4.8.5 Theoretical description of internal photoemission

Before proceeding with the experimental results of internal photoemission it may be interesting to take a look on the optical absorption properties of our sensors as predicted by the theory. For this purpose, we calculated absorption profiles, i.e. the depth dependence of the volumetric absorptance, as well as absorption

Table 4.5: Values derived from CV-curves in figure 4.60.

d_{ox} [nm]	Q_0 [$\frac{nC}{cm^2}$]	V_{FB} [V]	states [$\frac{1}{cm^2}$]	states [$\frac{1}{cm^3}$]	$\Delta\Phi$ [eV]
1	20	≈ 0 V	$1.25 \cdot 10^{11}$	$1.25 \cdot 10^{18}$	0.05
4	150	0.45 V	$9 \cdot 10^{11}$	$2.5 \cdot 10^{18}$	0.05

spectra, i.e. the wavelength dependence of the volumetric absorptance, in idealized Au(Pt)-SiO$_2$-Si sandwich structures according to reference [14,359]. Here, it is considered that not only the incident light but also reflection and interference effects between the incident and reflected waves contribute to the absorbtance in (nm-) thin films. In contrast, the absorption in the thick Si wafer can be and is treated by neglecting the latter two contributions, hence such that the simple Lambert-Beer law applies.

The volumetric absorptance $\eta_i(\lambda, x)$ is defined in layer i ($i = 1$ for Au(Pt), $i = 2$ for SiO$_2$, $i = 3$ for Si) and at a wavelength λ as the ratio of the number of photons absorbed per unit volume, dN/dV, at the position x into the sample from the surface, to the number of photons incident per unit area, N_0/S, of the irradiated surface:

$$\eta_i(\lambda, x) = \frac{1}{N_0/S} \frac{dN}{dV} \qquad (4.37)$$

and is calculated according to equation 14 of reference [359]:

$$\eta_i(\lambda, x) = \alpha_i \frac{\zeta_i'}{\zeta_0} [T(x,\lambda) + R(x,\lambda) + I(x,\lambda)], \qquad (4.38)$$

where T and R are the contributions from the transmitted and the reflected waves, while I is the contribution arising due to the interference of transmitted and reflected waves. The three contributions are calculated according to the relations:

$$T(x,\lambda) = |T_i|^2 \exp(-\alpha_i x) \qquad (4.39)$$

$$R(x,\lambda) = |R_i|^2 \exp(+\alpha_i x) \qquad (4.40)$$

and

$$I(x,\lambda) = 2Re\{(T_i^* R_i) \exp(-2i\zeta_i' x)\} \qquad (4.41)$$

4.8 Transport of excited holes through MIS sensors

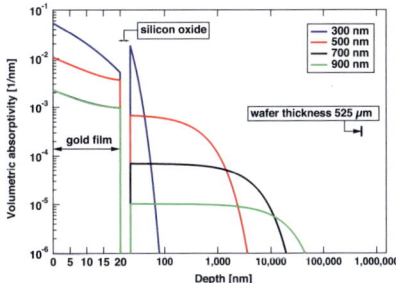

Figure 4.61: Calculated absorptivity profiles in a Si(525 μm)–SiO$_2$(3 nm)–Au(20 nm) sensor. Light incident on the gold top electrode.

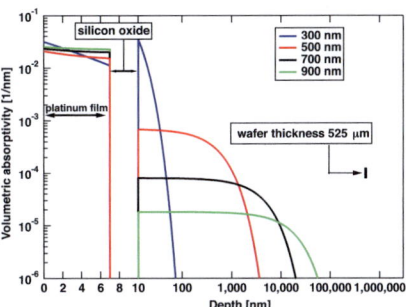

Figure 4.62: Calculated absorptivity profiles in a Si(525 μm)–SiO$_2$(3 nm)–Pt(7 nm) sensor. Light incident on the platinum top electrode.

where $\zeta_0 = 2\pi/\lambda$, $\zeta' = \zeta_0 n_i$, and $\alpha_i = 2\zeta'' = 2\zeta_0 k_i$, with n_i, k_i being the real and imaginary parts of the complex refractive index, respectively. T_i, R_i are the complex transmission and reflection coefficients in layer i, respectively. By solving a system of equations derived from applying the Maxwell equations for each layer and the Fresnel equations for each interface, one obtains an analytic expression for the depth dependence of the volumetric absorptance in each layer i (see reference [359]).

4.8.5.1 Absorption profiles

Figure 4.61 illustrates the absorption profiles calculated (as explained above) for a Si–SiO$_2$–Au layer system with thicknesses of 525 μm (Si), 3 nm (SiO$_2$) and 20

4 *Measurements, results and discussion*

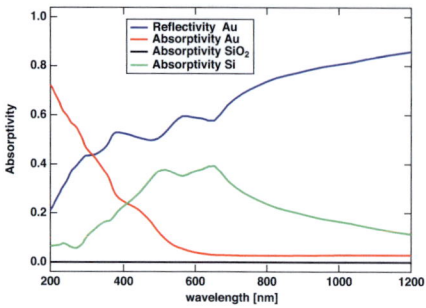

Figure 4.63: Fraction of incident photons absorbed in each layer of a Si(525 μm)–SiO$_2$(3 nm)–Au(20 nm) sensor.

nm (Au), respectively, for wavelengths of 300 nm, 500 nm, 700 nm, and 900 nm. The curves show a clear decrease of the absorptance with increasing wavelength in both the Au and the Si layer. Hereby, the absorptance in Si is by 1-2 orders of magnitude smaller than in Au when comparing their values at the two interfaces introduced by the oxide layer, except at 300 nm, where the absorptance in Si is slightly higher. The fact that Si is strongly absorbing at 300 nm is indicated also by the relatively steep absorption profile (note that the profiles in Si are represented on a logarithmic depth scale) which suggests that the photons entering the Si layer are all absorbed within the first 100 nm. At higher wavelengths, the profiles in Si are almost flat in the first 100 nm, indicating a weak absorptance in this region.

4.8.5.2 Wavelength dependence

By calculating the total absorptance in each layer for many wavelengths between 200 nm and 1200 nm, one can obtain the fraction N/N_0 of incident photons absorbed in each layer as a function of wavelength. For the Au system the wavelength dependence is depicted in figure 4.63. In the UV-region, most of the incident photons are absorbed in the Au layer. The corresponding fraction N/N_0 decreases, however, rapidly with increasing wavelength and drops, at wavelengths above 600 nm, below 4%. In contrast, the total absorptance in Si increases in the UV region and peaks at about 550 nm and 650 nm, but decreases in the IR region, where most of the incident photons are reflected at the Au surface, considered to be perfectly flat in our calculations.

A qualitatively similar behavior can be observed in the UV region of the total

4.8 Transport of excited holes through MIS sensors

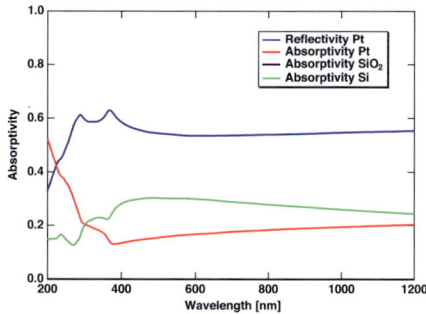

Figure 4.64: Fraction of incident photons absorbed in each layer of a Si(525 µm)–SiO$_2$(3 nm)–Pt(7 nm) sensor.

absorptance calculated for the Pt-based system, as shown in figure 4.64. While the total absorptance in Si exhibits again a maximum, this time at about 450 nm, followed by a slight decrease towards larger wavelengths, the absorbed fraction of photons in Pt does not show the monotonous decrease found in Au, but exhibits, instead, a minimum at about 400 nm, as well as a slight increase towards larger wavelengths. This latter behavior is apparently facilitated by the relatively poor reflectivity of the Pt surface compared to the one of the Au surface.

In conclusion, the Pt sensors show a significant absorptivity (around 20 % between 600 nm and 1200 nm) in both, the platinum and the silicon layer. Thus, photo excitation has to be considered in the silicon and the platinum layer. The gold sensors show only a weak absorptivity in the gold top electrode between 600 nm and 1200 nm. Thus, photo excitation will mainly occur in the silicon. This eases the discussion of the experimental results when both systems platinum and gold are compared. Hence, the experiments with internal photoemission are presented for systems with gold top electrode first and then compared to results for systems with platinum top electrode.

4.8.6 Experimental results

4.8.6.1 Internal photoemission without bias voltage

In figure 4.65, the internal photoemission yield, induced in each of the four sensors of the MIS sensor, is plotted on a i) linear scale (upper graph) ii) logarithmic scale (middle graph) as a function of the wavelength of the incident light.

In the upper graph, the calculated absorptivity of the silicon is added for com-

4 *Measurements, results and discussion*

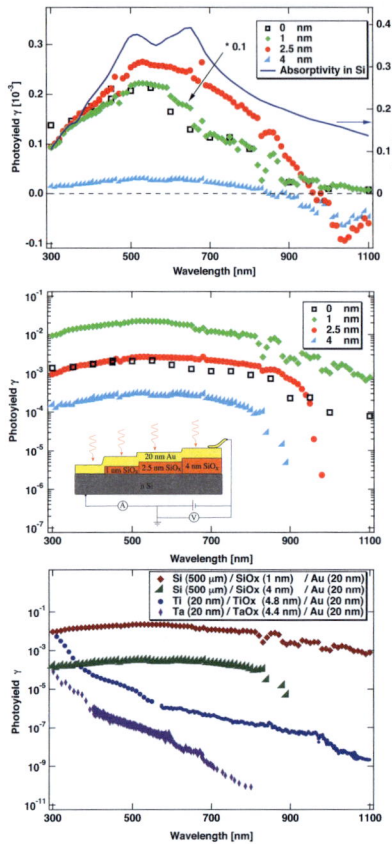

Figure 4.65: **Upper graph:** Linear plot of the internal photoemission yield as a function of the wavelength of the incident light as determined for the four sensors of the MIS sensor sketched in the inset of figure 4.67. The photoyield measured on the 1 nm oxide area of the MIS sensor is multiplied by 0.1 to enable a better comparison. The right axis shows the silicon absorptivity values from figure 4.63 which are plotted here as well. **Middle graph:** Logarithmic plot of the internal photoemission yield as a function of the wavelength of the light incident on the different areas of the sensor. Inset shows the set-up of the sample illuminated on the differently oxidized terraces of the silicon wafer. **Bottom graph:** Comparison of the photo yields for the MIS sensors (1 and 4 nm oxide thickness) with the photo yields of MIM systems (titanium and tantalum based) with gold top electrode of same thickness (taken from literature [83], compare section 4.6).

4.8 Transport of excited holes through MIS sensors

parison. Obviously the photoyield from 300 nm to 800 nm shows a similar spectrum. It was derived above that the photo excitation occurs mainly in the silicon. This is confirmed by the comparison of the photoyield and the absorptivity in the silicon. Both, the photoyield and the absorptivity show a maximum between 500 and 700 nm. In the mentioned wavelength range a photo induced electron flow is detected from the top gold to the bottom silicon electrode (an electron current from the top to the back is counted as positive photoyield throughout this work). This means one observes an electron current from the top to the back induced by a photo absorption in the back electrode. Thus, the excitation and transport mechanism for these photon energies should be described by a hole excitation in the space charge layer of the silicon and a subsequent neutralization of the hole from occupied conduction band states in the gold (denoted as processes 1 and 2 in figure 4.56).

For wavelengths from 800 nm to 1100 nm the absorptivity in silicon monotonously decreases. The same behavior is observed for photo yields studied with the sensors having 0 and 1 nm oxide thicknesses. A different phenomenon which cannot be explained by the absorptivity in the silicon layer appears for the sensors with 2.5 nm and 4 nm oxide thickness. The photoyield changes sign and becomes negative with considerable values of about 10^{-4}. The onset of this negative photoyield shifts with increasing oxide thickness toward lower wavelengths or higher photon energies. As mentioned before, the absorptivity in the silicon as well as in the gold is independent of the oxide thickness (no absorption in the band gap due to the assumed bulk properties of the thin oxide film). Thus, one can think about the appearance of different excitation mechanisms with increasing oxide thicknesses. Based on the capacitance voltage plots, I already derived above that thicker oxide layers contain disproportionate amounts of interface states (see table 4.5). An increased photon electron generation at the silicon–silicon oxide interface [360] due to the disproportionate occurrence of interface states would built up a source for photo electrons also at lower excitation energies. The interface states would allow an excitation of electrons by photons with $h \cdot \nu < E_{\text{gap}}(\text{Si})$, $E_{\text{gap}}(\text{SiOx})$ in the close vicinity of the gold top electrode. These electrons can easily propagate to the gold electrode when lower photon energies are used for the excitation. This would explain the polarity change of the photoyield for higher oxide thicknesses. The propagation of electrons to the gold electrode could be hindered by a negative bias voltage applied on the gold electrode. Thus, this excitation and transport model can be validated by studying the bias voltage dependence of the photoyield. This will be presented in the next section 4.8.6.2.

The photoyield for wavelengths from 300 nm to 800 nm decreases with oxide thicknesses from 1 to 4 nm. The absolute yields measured on the "2.5 nm" sensor are observed to decrease by about two-, while on the "4 nm" sensor even by three orders of magnitude with respect to the "1 nm" sensor. We attribute this decrease to the increasing distance between the space charge layer and the interface states at the silicon–silicon oxide interface on the one hand and the silicon oxide–gold interface on the other hand. So, scattering processes hamper the transport of excited carriers through the oxide and thus diminish the yield.

Similar to previous studies [319, 361, 362] I observe that the "0 nm" (MS) sensor is less photo-sensitive (a factor of ten in our case) than the "1 nm" (MIS) sensor. So yields of $10^{-3} - 10^{-2}$ are measured for the "1 nm" sensor, compared to $10^{-4} - 10^{-3}$ for the "0 nm" sensor. This difference of 1 order of magnitude could be assigned from a simple viewpoint to photon-electron coupling at the interface states since their density increases with increasing oxide thickness as mentioned above. However, the assumption that different states contribute to the photo excitation when comparing the 0 nm oxide sensor and the 1 nm oxide sensor is not appropriate. The spectrum of the photo yield (including excitation and transport) for both sensors and the absorptivity in the silicon still are consistent over the whole wavelength range from 300 nm to 900 nm. Thus the mechanism of photo excitation has to be the same. Only the transport of excited carriers seems to be eased by an order of magnitude by an 1 nm thick oxide layer on top of the bare silicon. One could speculate about the changes of internal dipole layers and built-in fields due to the thin oxide. However, the application of a bias voltage which changes the influence of dipole layers and built-in fields will shed more light on theses phenomena.

To further underpin the dominant role of the photo excitation in the silicon and the transport of excited holes I compare the spectrum of the MIS photo yield with the previously shown photoyields for titanium–titanium oxide–gold and tantalum–tantalum oxide–gold systems (compare sections 4.7, 4.5 and 4.6) [81, 83, 231]. The gold top electrode is equally prepared with the same thickness for both MIM and MIS systems. The MIM systems show a steep decrease of the photoyield over 5 orders of magnitude with increasing wavelength. As shown before, the strong exponential dependence of the photo yield on the photon energy (e.g. a steep slope up to 2.0 eV for tantalum oxide systems) can be described with a transport process through heterosystems with an internal tunnel barrier (compare section 4.6 [83]). In contrast, the MIS systems show nearly no slope, i.e.

a more or less unvaried photoyield over a wide range of photon energies. This points to a transport process with no or only a very small internal barrier. The height of this barrier cannot be assigned with these data where photon energies down to 1.1 eV are used.

4.8.6.2 Internal photoemission under applied bias voltage

The influence of the bias voltage on the photoresponse on the whole internal photoemission spectra can be seen in figure 4.66 where the photoyield of a MIS sensor with a 2.5 nm thick oxide layer is plotted linearly (upper graph) and logarithmically (lower graph) for four different bias voltages applied on the gold electrode. The sensor with 2.5 nm thick oxide was chosen, since it showed a deviation of the spectrum of the photoyield from the spectrum of the absorptivity in the silicon for wavelengths from 800 to 1100 nm. These experiments want to elucidate whether photocurrents are sensitive to an bias voltage applied on the gold electrode in this wavelength range. The deviation between the spectrum of the photoyield and the absorptivity in the silicon was tentatively attributed to an excitation of states at the silicon–oxide interface. It can be indeed observed, that already by applying -0.5 V bias voltage on the gold electrode, the polarity change between 800 nm and 1100 nm - discussed above - disappears. The photoyield shows the same spectrum as the absorptivity in the silicon over the whole wavelength range. Hence, this bias voltage applied on the gold electrode cancels the influence of the interface states, which were introduced by the insertion of the oxide between gold and silicon. Obviously the bias voltage applied on the gold electrode induces these differences in the spectra. The same spectrum is also recorded with -1.5 V, the photoyield is increased overall by a factor of 2.

Deviations of the photoyield from the absorptivity in the silicon can be detected for higher photon energies ($\lambda < 400$ nm) with a bias voltage applied on the gold electrode of -2.1 V. Absorption bands in thin silicon oxide layers are known just in this energy range (5.0 eV [360]). Thus, direct excitation of the silicon oxide and a subsequent field strength supported transport of the photo electrons to the silicon is the most probable process here.

To check the previous conclusion, that the insertion of the oxide film only changes the dipole layers and built in fields, I study the bias voltage (applied on the gold electrode) dependence of the photoyield for 635 nm for all four oxide thicknesses. The shape of the bias voltage (applied on the gold electrode) dependence is similar for all four sensors. The photoyield shows a saturation at negative bias voltages on the gold electrode < -1.5 V. Starting from positive bias

4 Measurements, results and discussion

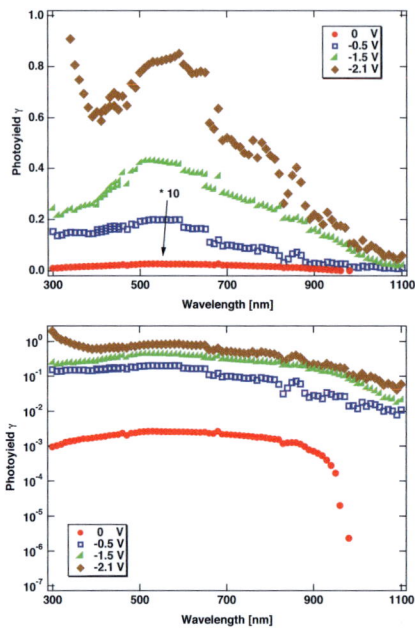

Figure 4.66: **Upper graph:** Linear plot of the internal photoemission yield as a function of the wavelength of the incident light for a MIS sensor with 2.5 nm thick silicon oxide while applying 0 V, -0.5 V, -1.5 V and -2.1 V bias voltage applied on the gold electrode. Values of the internal photoemission yield for 0 V bias voltage applied on the gold electrode are multiplied by a factor of 10. **Lower graph:** Logarithmic plot of the internal photoemission yield as a function of the wavelength of the incident light for a MIS sensor with 2.5 nm thick silicon oxide while applying 0 V, -0.5 V, -1.5 V and -2.1 V bias voltage applied on the gold electrode.

4.8 Transport of excited holes through MIS sensors

voltages of $< +0.5$ V on gold and tuning the bias voltage toward negative values the sensor with 1 nm thick oxide is the first to show an increase of the photoyield. Then the 0 nm, the 2.5 nm and the 4 nm sensors show the mentioned increases. This sequence is the same as the one for the maximum photoyield in figure 4.65. The samples with thickness from 1 to 4 nm thickness show the same saturation value for the photoyield. This supports our previous conclusion, that the insertion of oxide films with 1 to 4 nm oxide thickness only changes the dipole layers and built in fields. Additional scattering processes do not seem to be introduced, or they can at least be switched off by the application of a negative bias voltage on the gold electrode. This seems to be true, when one compares the 0 nm sample with the other three. The 0 nm sample shows a 20 % higher saturation value. We attribute this difference to transport losses by scattering processes in the oxide, which cannot be modified by the application of a bias voltage.

Starting from positive bias voltages of $> +0.5$ V on gold toward larger positive bias voltages, one can observe a local maximum of the photoyield around $+0.7$ V for the 0 nm oxide sensor. This value coincides with the height of the Schottky barrier usually existent in silicon–metal sensors [47,314,315]. For this bias voltage the Fermi level of gold and the valence band of silicon are roughly at the same level. Just in this case there is the highest efficiency to neutralize photo excited holes at the silicon–metal interface since the electronic density of states (DOS) in gold (s,p metal) is highest filled at the Fermi level. For more positive bias voltage applied on the gold electrode, the Fermi level of the gold starts to be below the valence band edge in the silicon. Therefore the neutralization efficiency for photo excited holes then decreases. It should be mentioned that for a positive bias voltage applied on the gold electrode the drift of excited holes in the space charge layer is weaker than with a negative bias voltage applied. Thus, the local maximum at $+0.7$ V exhibits a smaller value than the saturation value at negative bias voltages. The ratio for the 0 nm oxide sensor is $\gamma(-1.0\,\mathrm{V})/\gamma(+0.7\,\mathrm{V}) \approx 15$. The assumption, that at $+0.7$ V bias voltage the holes excited in the silicon are efficiently neutralized by Fermi electrons from the gold can be proven by the insertion of an oxide layer. Then no easy neutralization can happen. This is indeed the case. The local maximum for positive bias voltages on the gold electrode weakens considerably for 1nm thick oxide and shifts to $+0.2$ V and vanishes completely for higher oxide thicknesses. This finding supports the conclusion, that the photo current is driven by holes in the silicon even at positive bias voltages on the gold electrode.

Hole conduction from the silicon to the gold or electron transport vice versa has

4 Measurements, results and discussion

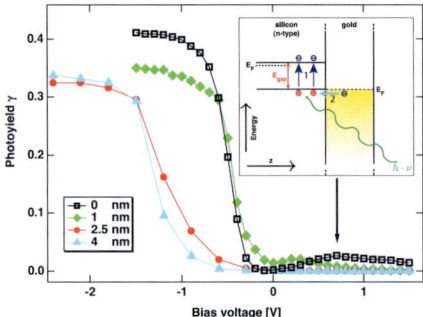

Figure 4.67: Internal photoemission yield as a function of the bias voltage induced by 635 nm radiation in each of the four terraces of the MIS sensor sketched in the inset of figure 4.65. Inset shows relative position of Fermi levels for the 0 nm oxide system at a bias voltage of $+0.7\,\text{V}$. Process 1 denotes the photo excitation in the silicon, process 2 is the neutralization of holes by electrons from the gold.

to pass through the oxide. The hole states in the silicon are in the midgap of the oxide. For amorphous oxide films there exists a number of midgap states [363]. Interesting for the hole conduction are states in the lower half of the gap. Such states are $\sigma - \pi$ states induced by the Si–O–O bond (4.2 eV above the valence band of the oxide). Also Si_3^0 (4.3 eV below the conduction band of the oxide) and Si_3^+ show highly localized states near the midgap [363].

Of course one could think about electron excitation in the gold and transport to the silicon for positive bias voltages on the gold electrode. This seems to be supported by the ratio of the absorptivities of silicon and gold, which is 12 for 635 nm (see also figure 4.63). This would nicely correspond to $\gamma(-1.0\,\text{V})/\gamma(+0.7\,\text{V}) \approx 15$. But this coincidence is misleading. In a later experiment the gold top electrode is exchanged with a stronger absorbing platinum electrode. An higher local maximum at positive bias voltages on the gold electrode should be expected then. But this is not the case (see figure 4.69). Hence, the contribution of photo excitation in the gold to the observed photocurrent is also be ruled out in this case.

A detailed overview over the bias voltage dependence for three example wavelengths for all four samples is shown in figure 4.68.

4.8 Transport of excited holes through MIS sensors

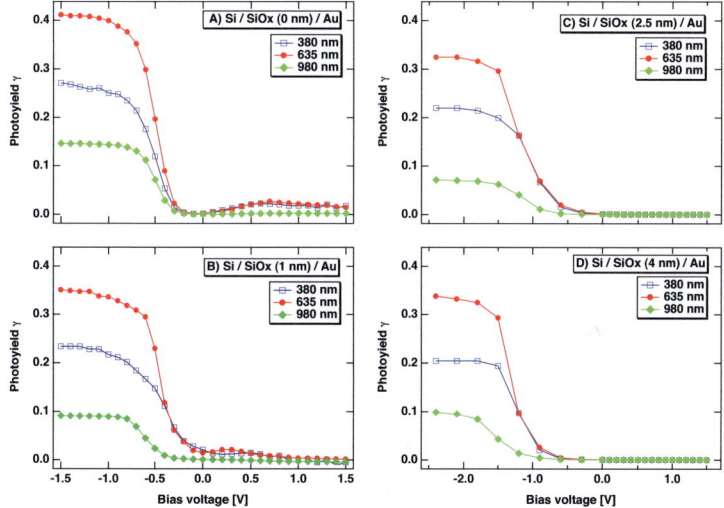

Figure 4.68: Internal photoemission yield as a function of the bias voltage induced by 380 nm, 635 nm and 980 nm radiation for the four steps of the MIS sensor: A) 0 nm, B) 1 nm, C) 2.5 nm and D) 4 nm.

4 Measurements, results and discussion

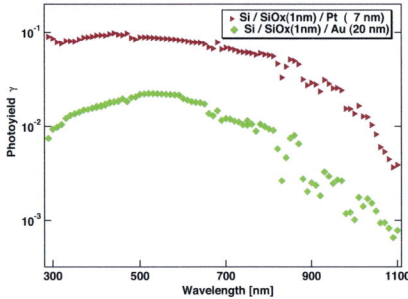

Figure 4.69: Logaritmic plot of the internal photoemission yield as a function of the wavelength of the incident light (on the 1 nm terrace) for a Pt and Au MIS sensor with 0 V applied bias voltage.

4.8.6.3 Internal photoemission with platinum sensors

As introduced above, the calculated absorptivities in the 7 nm thick platinum bias voltage applied on the gold electrode and the 20 nm thick gold electrode are roughly similar between 200 nm and 400 nm. But for wavelengths > 600 nm the absorptivity in the gold is very weak ($\approx$ 3 %), whereas a ten times larger value of 10 − 20 % is calculated for the platinum electrode (see figure 4.64). Despite this fact the observed spectrum of the photoyield is very similar for both sensors (see figure 4.69) and coincides over a wide wavelength range with the spectrum of the absorptivity in the silicon. This means that in platinum sensors the main photo excitation leading to a sensor current occurs also in the silicon.

Hence, the absorptivities in the silicon have to be compared (for example for medium wavelengths of around 600 nm). The absorptivity in the silicon is 0.3 in case of the platinum sensor and 0.4 for the gold sensor. This could point also to a comparable photoyield. However, the experimentally observed photoyield for the platinum sensor is around 8 times larger than for the gold sensor. Thus, a comparison of the sensors based on the absorptivities does not explain the measured yields.

The electrons in the gold electrode which neutralize the photo excited holes are mainly s,p electrons for low photon energies and d-electrons for higher photon energies (< 600 nm) [364]. For platinum d-electrons contribute to the photocurrent even for lower photon energies [365]. Additionally, the density of the electrons in the conduction band is higher. Therefore platinum supplies a higher electron density for the neutralization of the photo excited holes in the silicon,

4.8 Transport of excited holes through MIS sensors

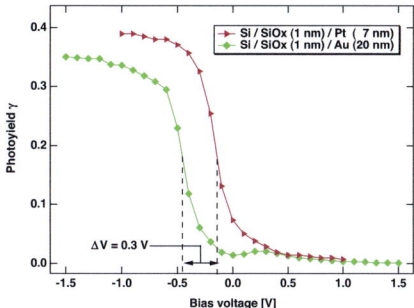

Figure 4.70: Internal photoemission yield as a function of the bias voltage induced by incident 635 nm light (on the 1 nm terrace) for a Pt and Au MIS sensor.

leading to a higher photoyield for the platinum sensors despite the comparable absorptivity in the silicon.

The higher photoyield for platinum sensors exists for all bias voltages. The bias voltage dependence for 635 nm wavelength is shown in figure 4.70 for a 1 nm oxide sample in comparison with the gold sensor with the same oxide thickness. Platinum sensors also show the same saturation behavior for negative bias voltages. When going to positive voltages the photoyields of the gold sensors decrease earlier than for the platinum sensors. The shift is $\Delta V = 0.3\,\text{V}$. At 0 V the photoyield for the gold sensors decreases already significantly whereas still 20 % of the photoyield can be observed for the platinum sensors. An argumentation using the DOS of the metals cannot explain the experimental result. The decrease of the photo yield for bias voltages $> -0.7\,\text{V}$ does not coincide with any significant change of the overlap of the photoexcited holes in silicon and the DOS structure of gold.

The DOS of the platinum shows a peak around E_F. Therefore one would expect a steeper decrease of the photoyield for positive bias voltages on the gold electrode for the platinum systems rather than for the gold systems.

Thus, the shape of the bias voltage dependence of the photoyield must be stronger influenced by the space charge layer in the silicon, the silicon–oxide interface and the electric fields in these layers. The shift of 0.3 V might be due to a change of the built-in electric field in the sensor.

4.8.6.4 Ion induced electronic excitations

In the previous sections we were able to show that the illumination with visible light leads predominantly to a photo excitation of the semiconductor. This section now deals with a selective electronic excitation of the top electrode only. For this purpose we use an argon ion beam of 200 eV kinetic energy. The primary ion current (penetration depth $\approx$ 1 nm, compare figure 3.3) impinging normal to the surface of the MIS sensor on the gold top electrode is 1 μA (compare process 5 in figure 4.56).

Excited holes and electrons in metals have a lifetime τ of some 10 fs [366]. Assuming the velocity of an excited hole v_h being equal to the Fermi velocity of an electron ($v_F \approx 10^6$ m/s [367]) such lifetimes lead to a mean free path $\lambda = v_F \cdot \tau$ of 10 nm. This value corresponds to our film thicknesses. Thus, in our experiment a ballistic transport of the ion induced electronic excitations to the inner interface of the sensor is still possible.

The ion beam is switched on between 40 s and 60 s and between 100 s and 120 s. Otherwise it is switched off. For a bias voltage of 0 V the sensor current follows the square wave trace of the primary ion beam impinging on the sample (see red curve in figure 4.71). The sensor current reaches a constant value of 200 nA immediately after switching on the beam. The polarity of the current corresponds to an electron flux from the silicon to the gold top electrode or a hole flux vice versa. With an impinging primary ion current of 1 μA the amplitude of the sensor current corresponds to a yield of -0.2. The yield varies only weakly from -0.16 to -0.22 for positive bias voltages between 0 and 1.0 V.

When a negative voltage is applied, a constant sensor current cannot be detected. Instead, transient shapes similar to charging and discharging currents in capacitors are observed, see blue curve in upper graph of figure 4.71. When the primary ion beam is switched on for a longer period, the induced current approaches zero, as the blue curve shows between 150 s and 200 s. The ion induced current is in this case not a steady state current since it decays always to zero after continuing ion beam exposure. In this case we assign a yield of 0. The bias voltage applied on the gold electrode dependence of the ion yield is shown in the lower graph of figure 4.71.

The polarity of the square wave trace for positive bias voltages on the gold electrode indicates a hole current from the gold to the silicon or an electron current vice versa meaning the same current direction as observed in the photoemission studies for wavelengths shorter than 800 nm. Considering the excitation process one would expect an electronic distribution which consists of excited electrons

4.8 Transport of excited holes through MIS sensors

and holes quite symmetric to the Fermi level [368]. But only a hole current can be observed. This means that the current contribution due to hot electrons in the top electrode is suppressed (see inset in the lower graph of figure 4.71). This surprising fact is consistent with the optical experiments reported before. In conclusion, both the photo current and the ion induced sensor current are hole currents. The photo current is due to excitation in the semiconductor and the ion induced sensor current is due to hole excitation in the metal. The excited holes in the metals can be neutralized by electrons from the semiconductor since at these positive bias voltages on the gold electrode, the semiconductor is in accumulation. This means that the space charge layer is flooded by the majority carriers, electrons in this case. On the other hand with negative bias voltages on the gold electrode, the space charge layer is depleted. Thus a steady electron current from the semiconductor neutralizing the excited holes in the metal does not exist. But the capacitor like charging and discharging currents can be observed.

These charging and discharging phenomena can be observed for all negative bias voltages. The maximum amplitudes of the charging and discharging traces of ≈ 200 nA correspond to the plateau current values which are observed for positive bias voltages.

As mentioned before $\sigma - \pi$ states in the amorphous oxide enable a conduction of the photo excited holes through the oxide. The holes excited by the ion beam may use the same transport channels. When the draining of these states is hindered by negative bias voltages on the gold electrode, these states would be charged. This leads to an increase of the internal electric field of the sensor which acts as a counter voltage for further carrier transport. The ion induced sensor current then dies out. This means that the bias voltage applied on the gold electrode does not change the transport channel used by the excited holes. But the bias voltage applied on the gold electrode modifies the draining of the states which are used.

The charge following during the "beam on" period is 1 to 2 μC (blue shaded areas). With an exposed area of 0.08 cm^2 one gets a charge density for the first "beam on" period of 25 μC/cm^2.

The charge during the first "beam off period" (gray shaded areas) is around 70 % larger than the charges flowing during the "beam on" periods (values of $\approx 4 \mu$C).

In the next experiment, the ion beam is again switched on when the ion induced discharging of the oxide is not completed (at $t \approx 100$ s). Then the subsequent ion induced charging is much smaller (from $t \approx 100$ s to $t \approx 130$ s). The charge is only

4 Measurements, results and discussion

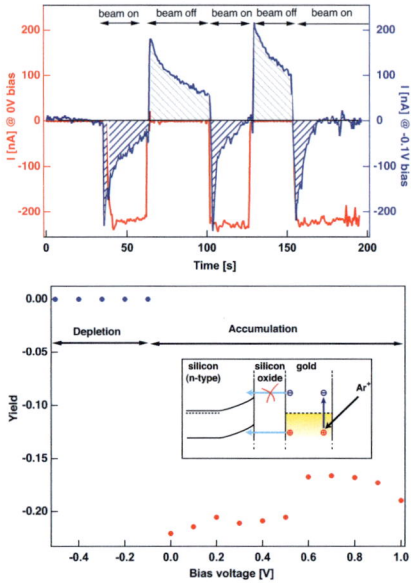

Figure 4.71: **Upper graph**: Transient shape of the ion induced sensor current I_{dev} for two bias voltages. Primary ion current impinging on the sensor surface $I_{inc} = 1\,\mu A$. **Lower graph**: Yield (ratio of steady state ion induced sensor current and primary ion current) as a function of the bias voltage.

4.8 Transport of excited holes through MIS sensors

half of the previous value (12 μC/cm^2). This reduced charging is again found in the next cycle (from $t \approx 150$ s to $t \approx 200$ s). A complete discharging of the oxide states can be reached by a long beam off time. Then a complete charging of 25 μC/cm^2 can be found again. This means that the excited hole induced states have a significant life time of several tens of seconds. Life times in that range can also be found by deep level transient spectroscopy for silicon oxide systems [369].

In general the charge when the ion beam is switched off (electrons from the gold to the silicon) is larger than the primary hot hole induced charge. Excited electron as well as excited hole states with high excess energy can easily decay in multiple carriers with lower excess energy. Such a conversion process may occur when the ion beam induced hole states drain after the beam is switched off.

The excited hole induced charge densities must be compared with the capacitances shown in figure 4.60. The ion induced excited holes charge the metal–oxide interface. The oxide dominated capacitance in the capacitance-voltage plot shows up in the accumulation region (positive bias voltages on the gold electrode) of figure 4.60. The value for the sensor with the 1 nm oxide film is $\approx$ 400 nF/cm^2. The ion induced charge for a complete charging is 25 μC/cm^2. This would correspond to an internal ion induced voltage of $\approx$ 60 V. But one has to keep in mind that the capacitance voltage plot probes the states with energies $E = E_F \pm U_{bias}$ whereas the energy of excited hole states can extend to the bottom of the gold conduction band. Thus, the ion beam induced holes have access to a broader window of the interface state distribution than the ground state carriers which charge the capacitance at a certain bias voltage.

If we assume singly charged states the value of 25 μC/cm^2 would correspond to $1.5 \cdot 10^{14}$ states cm^{-2}. With an atom density of $\approx 8 \cdot 10^{14}$ states cm^{-2} for the silicon [351] this would correspond to a maximum charging of approximately 20 %, for the case of singly charged states, of the interface.

In conclusion the dominant role of excited hole conduction and hole induced charging processes was derived by the comparison of the ion beam and the photo experiment. The hole conduction through the oxide can be assigned as a property of the oxide only and not to special interface states at the oxide–gold interface. The latter interpretation can be ruled out by exchanging the gold top electrode with a platinum top electrode, also in the ion beam experiment. Also in these experiments (not shown here) a hole current from the platinum to the silicon can be detected.

On platinum and gold sensors the dependence of the ion induced yield on the spectral distribution of the electronic excitation is studied. The spectral distribu-

4 Measurements, results and discussion

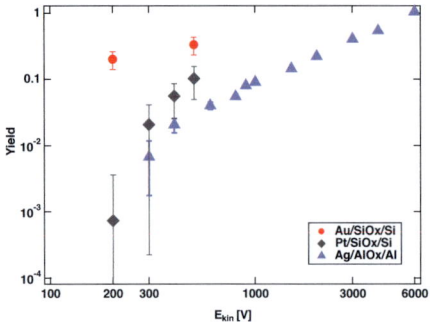

Figure 4.72: Ar$^+$ ion induced signals, logarithmically plotted as a yield, in dependence of the incident beam energy recorded with silicon(500 μm)–silicon oxide(1 nm)–gold(20 nm), silicon(500 μm)–silicon oxide(1 nm)–platinum(7 nm) MIS and aluminum(50 nm)–aluminum oxide(4 nm)–silver(15 nm) MIM sensors (taken from [15]). All yields taken at 0 V bias voltage.

tion can be controlled by the kinetic energy of the impinging ion beam [370, 371]. For platinum sensors the yield is about two orders of magnitudes smaller at $E_{kin} = 200$ eV, for $E_{kin} = 500$ eV the difference is only a factor of 3. The lower sensitivity for platinum sensors at low kinetic energies is due to the stronger electron-electron interaction in platinum compared to gold. This results in to a strong damping of weak electronic excitations (also found in ballistic electron emission microscopy [372]). At higher kinetic energies the penetration depth of the argon ions is larger, thus the transport length for excited electrons decreases and damping becomes less important.

For comparison the ion induced electron yields observed with an aluminum (50 nm)–aluminum oxide (4 nm)–silver(15 nm) **m**etal–**i**nsulator–**m**etal sensor (MIM) are added in figure 4.72. It should be mentioned, that the MIM-systems at 0 V bias voltage are preferentially conductors for excited electrons, whereas the here presented MIS-sensors are preferential hole conductors.

The MIM systems show a lower yield than both MIS systems. When one assumes comparable transport losses in the 15 nm thick silver film of the MIM and the 20 nm thick gold film of the MIS system one can assign the higher sensitivity of the MIS compared to MIM systems to a lower internal barrier. This agrees with the different spectrum of the photoyield as mentioned above.

4.8.7 Conclusion

I have shown that MIS systems can act as very sensitive hot hole detectors. The conduction of excited holes predominates over the hot electron conduction. Additionally the hot hole conductivity is switchable by an applied bias voltage. The switching behavior can be modified by a thickness variation of the interstitial oxide layer (by the localized chemical oxidation of different areas on one wafer), enabling also an easy fabrication of sensor arrays. The spectral analysis of the internal photoemission yield shows that the transport of excited carriers with excess energies as small as 1.1 eV is not hindered by an internal barrier. The sensor acts as detector with broadband sensitivity.

The high sensitivity found in internal photoemission studies (measured with negative bias voltage) is caused by the efficient collection of hot holes in the space charge layer of the silicon substrate (holes from silicon to top electrode). The high sensitivity in the ion induced internal electron emission experiment is due to the effective hole conduction (holes from the top electrode to silicon). In the photo and the ion experiment the yields are comparable (for gold sensors $\approx$ 0.3 in reverse biasing for the photons and $\approx$ 0.2 in forward biasing for the ions).

This makes the sensors valuable for the study of surface reactions, where a strongly negative adsorbate pulls an electron out of the surface. Such electronegative reactions might be for example the F_2^- formation via the reaction between a low energy (30 - 300 eV) F^+ ion beam and an fluorinated silver or a silicon surface [373, 374] or O_2^- formation via the interaction of a low energy O^+ ion beam (10 - 60 eV) [375].

By this way the same kind of surface excitations as presented with Ar^+ can be studied with the help of MIS sensors integrating the reaction surface as the top electrode.

Furthermore one can think of an application of the MIS sensors as very sensitive temperature sensors. Due to the fact that the MIS sensors exhibit such a small internal barrier, it should be very sensitive for smallest temperature changes.

4.8.8 Mathematical Appendix

This appendix shows how an ideal CV-curve can be derived by a mathematical algorithm from experimental CV-curves. The total oxide charge Q_{ox} can be determined by equation 4.35.

The ideal CV-curve has then to fulfill the following equations:

$$\int C_1(V)\,dV - \int C_{ideal}(V)\,dV = Q_{ox}/2 \qquad (4.42)$$
$$\int C_{ideal}(V)\,dV - \int C_2(V)\,dV = Q_{ox}/2$$

Therefore we set-up a list of C-values $C[i]$ ranging from C_{min} to C_{max} with 160 elements. Then for every value $C[i]$ the solutions of the equations

$$\begin{aligned} C_1(V_1[i]) &= C[i] \\ C_2(V_2[i]) &= C[i] \end{aligned} \qquad (4.43)$$

can be found (compare horizontal line and denotation in figure 4.73). Thereby the ideal voltage value $V[i]$ is the arithmetic mean:

$$V[i] = \frac{V_1[i] + V_2[i]}{2}. \qquad (4.44)$$

The ideal curve $C_{ideal}(V[i])$ is then plotted versus these mean bias voltage values, resulting in the black dashed curve in figure 4.73. This algorithm was checked to fulfill the above mentioned equations 4.43.

4.8 Transport of excited holes through MIS sensors

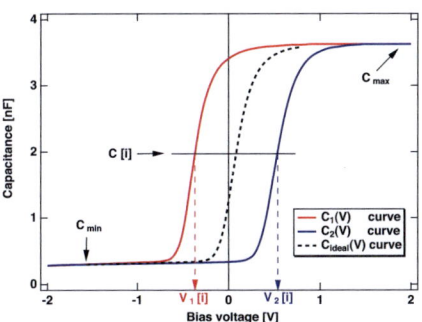

Figure 4.73: Determination of an ideal capacitance voltage curve $C_{\text{ideal}}(V)$ from experimental capacitance voltage curves $C_1(V)$ and $C_2(V)$.

4.9 Molecular reaction chemicurrent studies with MIS sensors

4.9.1 Study of effusing molecular gases with MIS sensors

The simplest way to study molecular reactions with the sensors is to dose just some molecules onto its surface. For a more advanced study of other molecular reactions, such as the water formation reaction on the surface of the chemicurrent sensor (see figure 4.74), a mixture of oxygen and hydrogen can be dosed via a gas inlet system or directed by a molecular beam onto the surface of a chemicurrent sensor. The part of the energy dissipated into the electronic excitation of the substrate can then be measured as a sensor current [10, 45]. The expected amount for the overall released energy during the water formation reaction on platinum and subsequent water desorption is shown in figure 4.75 for a reaction temperature in the range, where a desorption of water occurs with a significant rate (> 170 K) [98]. With these information one should be able to determine how much of the released energy (-71 kJ $\cdot$ mol^{-1} when H_2O desorption takes place after its formation [376, 377]) is transferred into the electronic excitation [10, 45] in the final step of the reaction. This assumption is especially valid if the product molecule does not carry a significant amount of energy away, as it is valid for the water formation reaction [45, 378].

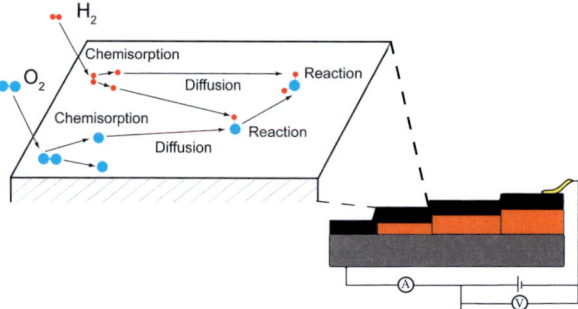

Figure 4.74: Schematic drawing of the water formation reaction which is carried out on the surface of a chemicurrent sensor.

The most sensitive sensors should be used to measure these chemicurrents and record the smallest possible signals. We recently published a calculation concerning the sensitivity of different chemicurrent sensors [379]. We showed a model

4.9 Molecular reaction chemicurrent studies with MIS sensors

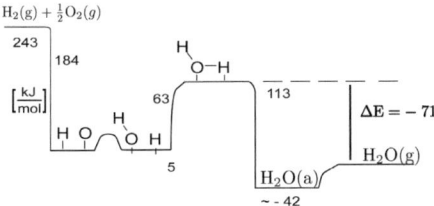

Figure 4.75: Potential energy diagram for the water formation on platinum for a reaction temperature $T_{\text{reaction}} > T_{\text{desorption}} = 170\,\text{K}$ [98] according to [376, 377], energies denoted in kJ/mol.

calculation for a metal which is moderately heated ($\Delta T = 10^{-4}\,\text{K}$) and separated via an e.g. 1 eV high barrier for electrons from a second metal, respectively semiconductor layer, the temperature of which is kept constant. This temperature difference across the device leads to an electron current from the warmer to the colder electrode. The silicon based systems showed the largest current densities, e.g. we found a value of $10\,\text{nA} \cdot \text{cm}^{-2}$ for 1 eV barrier height. Thus, one would choose the MIS system to measure the chemicurrents. This can be underpinned further by the photoemission studies presented before. MIS systems with gold, respectively platinum top layer, show even higher sensitivities in internal photoemission spectra studies than titanium based sensors (compare sections 4.8 and 4.6, [83]). Therefore, the MIS systems with the catalytic active platinum top electrode will be the working horse in the molecular chemicurrent experiments. All experiments are carried out on the 1 nm thick oxide area of a Si–SiOx(1 nm / 4 nm)–Pt sensor at ambient temperatures (at 0.01 V bias voltage applied on the gold electrode to ease a recordable signal, compare section 4.8). However, one has to keep in mind, that the systems with the low barriers (as the MIS systems) show stronger thermoelectric effects (e.g. the mentioned $10\,\text{nA} \cdot \text{cm}^{-2}$ for a temperature difference of $\Delta T = 10^{-4}\,\text{K}$) for all kind of excited electrons, as well. Therefore, one should always perform crosscheck experiments regarding the photosensitivity (compare sections 4.7 and 4.8) and the temperature dependence of the e.g. photosensitivity (see section 6.2), when chemicurrents are studied.

Before the experiments, the platinum top surface has to be cleaned. One way is to use low energy (90 eV) Ar^+ or H_2^+ ions to sputter the surface for several 10 minutes with a few 100 nA ion beam current.

4 Measurements, results and discussion

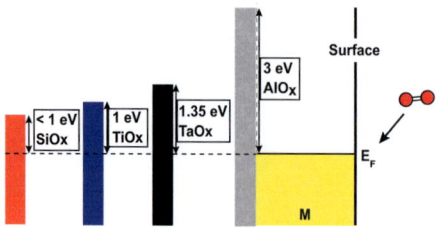

Figure 4.76: Hetero sensors with different mean barrier heights (MIM and MIS sensors). Typical mean barrier heights for AlOx [14], TaOx [85] (compare section 4.6, [83]), TiOx (compare section 4.6, [83]), and electrochemically prepared SiOx (compare section 4.8) are shown [379].

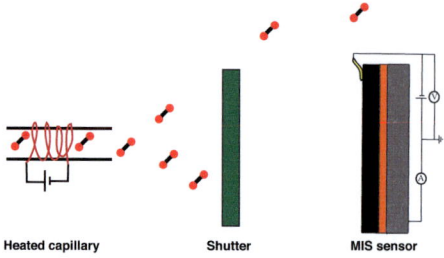

Figure 4.77: Drawing of the set-up used for a direct dosage of excited species. A mechanical shutter (tantalum flag) can interrupt the direct dosage of the incident species effusing through the heated capillary to the MIS sensor.

4.9.2 Direct dosage of excited species on a MIS sensor I

When the surface is cleaned, a current can be recorded with the platinum MIS sensor when gases such as hydrogen or oxygen are dosed into the chamber using a heated capillary. Whereas a dosage of argon does not give a measurable current signal.

First, I present the observed chemicurrents when a direct dosage with excited hydrogen species is carried out. In this set-up the platinum MIS sensor is in line with a heatable capillary. A mechanical shutter (tantalum flag) is used to interrupt the direct dosage of the incident species effusing through the heated capillary to the MIS sensor. When the shutter is closed, the species have to undergo several collisions with the chamber walls before reaching the MIS sensor (see figure 4.77).

Chemicurrents of hydrogen observed with the platinum MIS sensor are shown in figure 4.78. For these experiments the capillary is heated with 8 A and 7.45 V to a temperature of 1700 K and fed with molecular hydrogen ($p_{feed} = 7 \cdot 10^{-3}$ mbar $- 9.9$ mbar, $p_{chamber} = 10^{-4} - 10^{-7}$ mbar). The used temperature is clearly below the conditions at which formation of a significant amount of atomic hydrogen ($< 10^{-3}$ dissociation degree) occurs [7–9, 42, 104, 105, 380], compare as well the later following discussion in section 4.9.4. In addition no ions can be formed at the used temperature of 1700 K. The absence of ions in the expanding beam was verified by measuring the current between the platinum top electrode and ground.

In figure 4.78 the dependence of the recorded signals on the gas pressure in the chamber are shown. As a second x-axis the corresponding particle flux coming out of the heated capillary is introduced. These particle fluxes were determined taking the pressure in the UHV chamber, the pumping speed and the gas specific ionization factors [381] into account. The particle flux per second (N/s) flowing into the chamber can be calculated via formula 4.45, based on the gas pressure $p_{chamber}$ in the chamber (after scaling with the ionization factors) [106].

$$\frac{N}{s} = \frac{p_{chamber} + 2.637 \cdot 10^{-8} \, \text{mbar}}{7.824 \cdot 10^{-23} \, \text{mbar} \cdot \text{s}} \quad (4.45)$$

The recorded signals are negative, meaning a current is flowing from the bottom silicon to the top platinum electrode. This is the same current direction as it was found in the experiments where the MIS sensors were exposed to Ar^+ ions (presented in section 4.8). In that section it was shown that the MIS sensors are very effective excited hole conductors. Since the samples were identically prepared, the same conduction mechanism for excited holes may be operative. Thus, an excitation of electrons and holes in the top platinum electrode may result in an effective hole current from the front platinum to the back silicon electrode (respectively to the neutralization electron current from the silicon to the platinum).

The recorded signals increase with higher particle fluxes to larger negative values. It seems to be that there is always the same difference between the signals with opened and closed shutter. In the shutter open position photons and particles impinge directly on the MIS sensor surface. In the shutter closed position reflected light and particles can reach the MIS sensor surface after (several) collisions with the chamber walls. Thus this constant difference of the signals can be interpreted as a photo induced signal of ≈ 50 nA. However, it is very interesting to note, that there is always just this constant difference between the signal with

4 Measurements, results and discussion

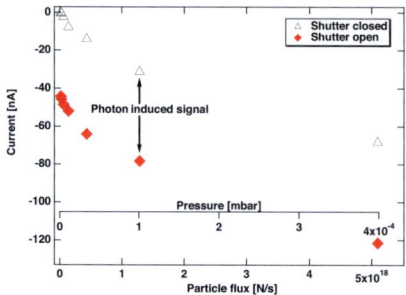

Figure 4.78: Hydrogen is heated via a capillary to 1700 K. The chemicurrent measured with a platinum MIS in dependence of different fluxes is shown. When the shutter is in the open position, photons and particles impinge directly on the MIS sensor surface. When the shutter is in the closed position, reflected light and particles can reach the MIS sensor after several collisions with the chamber walls. Chemicurrents are plotted versus the pressure in the chamber, respectively the particle numbers per second coming into the chamber. Data are shown for +0.01 V bias voltage applied on the gold electrode.

opened and with closed shutter. Meaning, apart from the photo induced signal, no difference between direct or indirect (after several collisions with the walls) impinging of hydrogen molecules (heated via the capillary) can be observed.

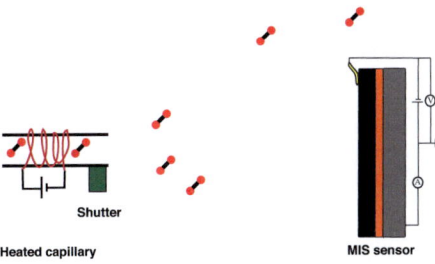

Figure 4.79: Drawing of the set-up used for a direct dosage of excited species with a capillary equipped with an internal shutter (shown for open position). This mechanical shutter (tantalum flag) can interrupt the direct dosage of the incident molecules effusing through the heated capillary to the MIS sensor in the source.

4.9.3 Direct dosage of excited species on a MIS sensor II

In addition to the experiment with the heated capillary presented in section 4.9.2 a second capillary set-up is used. This set-up is sketched in figure 4.79. As in the previously mentioned experiment the platinum MIS sensor is in line with the heated capillary. But in this experiment the mechanical shutter (tantalum flag), which is used to interrupt the direct dosage of the incident molecules effusing through the heated capillary to the MIS sensor, can close the capillary in the front. Therefore no photons from the heated capillary can reach the MIS sensor when the shutter is closed. Particles have to undergo several collisions with the shutter and the capillary to exit the heated capillary source anywhere, but not in line with the MIS.

The chemicurrents observed with a platinum MIS sensor in this set-up are shown in figure 4.80. Here the capillary is heated to temperatures between 1320 K and 2160 K and fed with $p_\text{feed} = 1\,\text{bar}$ ($p_\text{chamber} \approx 10^{-6}\,\text{mbar}$) molecular hydrogen, respectively oxygen. As before no ions are observed when the current between the platinum top electrode and the ground is probed.

The measured chemicurrents (left y-axis) and photon induced currents (where the same parameters are used, just without any gas flux, right y-axis) are plotted in the upper graph of figure 4.80 versus the heated capillary temperature. The measured photon induced currents are much larger than the chemicurrents (compare figure 4.80). All signals increase with higher capillary temperatures. When the shutter is closed no photon or particle induced signal is measured with the platinum MIS sensor. More light can be shed on these data when they are plotted in an Arrhenius plot. This Arrhenius plot, where the natural logarithm of the measured currents is plotted in dependence of the inverse temperature, can be found in the lower graph of figure 4.80. From this, one can obtain an activation energy of 1 eV for the processes leading to the signals in the experiment with oxygen and hydrogen feeding of the capillary via the linear slope m_{slope} and the Boltzmann constant R ($E_a = -R \cdot m_{slope}$). For the crosscheck experiment without any gas flux, where just photon induced currents are observed, an activation energy of 1.5 eV is obtained. Thus one can state that the processes leading to the signals in the experiment with the gases are less influenced by the temperature than the photon induced currents. The temperature dependence of the photon induced currents in chemicurrent sensors will be presented later in sections 6.1 and 6.2. From these studies, and the difference in the activation energies between the chemistry and the photon related signals, one can exclude the sensors characteristic response on the temperature as the origin of the observed temperature dependence chemicurrents.

Furthermore, due to the comparable signals for oxygen and hydrogen gas feeding of the capillary, one can exclude a migration of hydrogen through the platinum, as it would be expected for palladium surfaces [382–384]. When such a migration of hydrogen through the platinum layer to the silicon oxide is assumed as origin of the signal, one would expect a much larger signal for hydrogen than for oxygen. But that is not the case here.

4.9.4 The origin of the chemicurrent signals

In the following I would like to discuss different kinds of probable excitation processes due to the heated capillary in terms of:

- Ions
- Metastable molecules
- Atoms
- Vibration of molecules

In the experiment with the heated capillary just voltages smaller than 15 V are used and no ions impinging on the sensor's surface are detected.

Of course one could think of some metastable states of the particles that can be populated by the heated capillary. However, such metastable states would not

4.9 Molecular reaction chemicurrent studies with MIS sensors

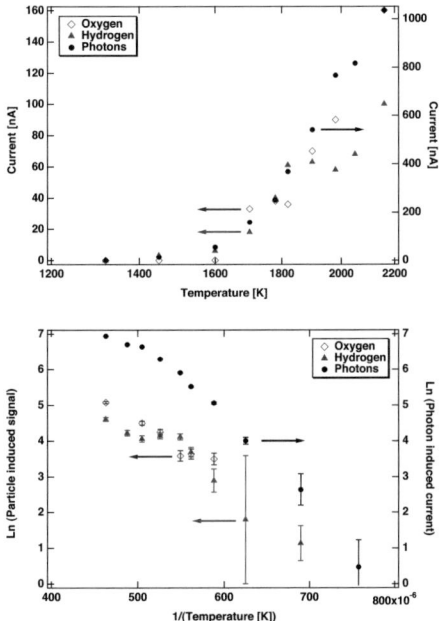

Figure 4.80: Chemicurrents and photon induced signals obtained in an experiment sketched in figure 4.77. The **upper graph** shows the measured currents in dependence of the heated capillary temperature. The **lower graph** shows the natural logarithm of the measured currents in dependence of the inverse temperature, in an Arrhenius plot.

exist anymore past collisions. But in the experiment with the shutter placed in a distance from the capillary, a chemicurrent signal increase was found even when the shutter (in distance, not closing the capillary) was in its 'closed' position (what can not be attributed to a photon induced signal). Meaning the particles had to pass around the shutter via collisions with the walls to reach the sensor's surface. Furthermore, in a later section 4.9.6 an experiment is presented where an indirect dosage of excited species is carried out and chemicurrents are measurable. In this experiment the species have to undergo several collisions to reach the sensor's surface. Thus I exclude metastable states as the origin of the measured chemicurrents.

Already Langmuir [16–19] described that molecular hydrogen dissociates when a hot tungsten wire is present. Following the literature of Langmuir [17–19] and the calculations by Bonhoeffer (based on Langmuir's data) [385, 386] one can obtain with the equilibrium constant $K(T)_{hydrogen}$ the fractional dissociation $\alpha_{hydrogen}$ for hydrogen by the following equations:

$$K(T)_{hydrogen} = 10^{\frac{-21200}{T} + 1.765 \cdot \log(T) - 9.85 \cdot 10^{-5} \cdot T - 0.265} \tag{4.46}$$

$$\frac{(\alpha_{hydrogen})^2}{1 - \alpha_{hydrogen}} = \frac{K(T)_{hydrogen}}{p} \tag{4.47}$$

When equations 4.46, 4.47 are then solved for the pressure p in the capillary heated to the temperature T, one obtains the fractional dissociation $\alpha_{hydrogen}$ for hydrogen. This dissociated fraction of H atoms is shown in the upper graph of figure 4.81 in dependence of the feeding pressure (p) with molecular hydrogen for temperatures between 1500 K and 2500 K of the capillary. One can clearly see that with lower feeding pressures and higher temperatures more hydrogen molecules are dissociated to hydrogen atoms (larger fractional dissociation $\alpha_{hydrogen}$).

$\Theta_{rot}(O_2)$	$\Theta_{vib}(O_2)$	m_{proton}	ΔD_0	$g_e(O)$	$g_e(O_2)$
2.08 K [387]	2274 K [388]	$1.67 \cdot 10^{-27}$ kg	5.116 eV [389]	9 [390]	1 [390]

Table 4.6: Literature values needed for the calculation of the fractional dissociation α_{oxygen} of oxygen according to equations 4.48, 4.49, 4.50 and 4.47 [391].

An analogous graph for the fractional dissociation α_{oxygen} of oxygen can be obtained using the rotational temperature, $\Theta_{rot}(O_2)$, the vibrational temperature $\Theta_{vib}(O_2)$, the proton mass m_{proton}, the difference of the dissociation energies ΔD_0, the degeneracy factors ($g_e(O)$, $g_e(O_2)$ in accessible energy range), which are listed in table 4.6. With these data, the molar masses ($M(O)$, $M(O_2)$), the Boltzmann constant k_B and the Planck constant h, the fractional dissociation α_{oxygen} of oxygen can be calculated in the following way [392]:

The quotient of the particle partition function $z(O)$, $z(O_2)$ and the volume V is:

$$\frac{z(O)}{V} = \left(\frac{2 \cdot \pi \cdot m_{proton} \cdot M(O) \cdot k_B \cdot T}{h^2}\right)^{\frac{3}{2}} \cdot g_e(O) \quad (4.48)$$

$$\frac{z(O_2)}{V} = \left(\frac{2 \cdot \pi \cdot m_{proton} \cdot M(O_2) \cdot k_B \cdot T}{h^2}\right)^{\frac{3}{2}} \cdot \frac{T}{2 \cdot \Theta_{rot}(O_2)} \cdot \frac{1}{1 - e^{-\frac{\Theta_{vib}(O_2)}{T}}} \cdot g_e(O_2) \quad (4.49)$$

Then the equilibrium constant $K(T)_{oxygen}$ for oxygen can be obtained by the following equation [391]:

$$K(T)_{oxygen} = \frac{(\frac{z(O)}{V})^2}{(\frac{z(O_2)}{V})} \cdot k_B \cdot T \cdot e^{-\frac{e \cdot \Delta D_0}{k_B \cdot T}} \quad (4.50)$$

The fractional dissociation α_{oxygen} of oxygen can now be obtained, using this equilibrium constant $K(T)_{oxygen}$, by solving equation 4.47 [7]. The results are plotted in the lower graph of figure 4.81 in dependence of the feeding pressure (p) with molecular oxygen for temperatures between 1500 K and 2500 K of the capillary. It can be seen that with lower feeding pressure and higher temperature more oxygen molecules are dissociated to oxygen atoms (larger fractional dissociation α_{oxygen}). However, the dissociated fraction α_{oxygen} is smaller than the dissociated fraction $\alpha_{hydrogen}$ for the same pressures and temperatures.

Later in this section these dissociated fractions $\alpha_{hydrogen}$ and α_{oxygen} are used to obtain some information about how many molecules are dissociated, and how many molecules are in the first excited state to get more insight into the in section 4.9.3 described experiment.

Therefore, I would like to shed some light on vibrationally excited molecules. However, the life times in the free gas (not adsorbed on a surface) of such vibrationally excited molecules is not easy to determine. But one can estimate that they are at least a few milliseconds to seconds [393,394]. Furthermore some stud-

---

[7] Here one should not use the equations presented in [392], since there is an error in the book.

4 Measurements, results and discussion

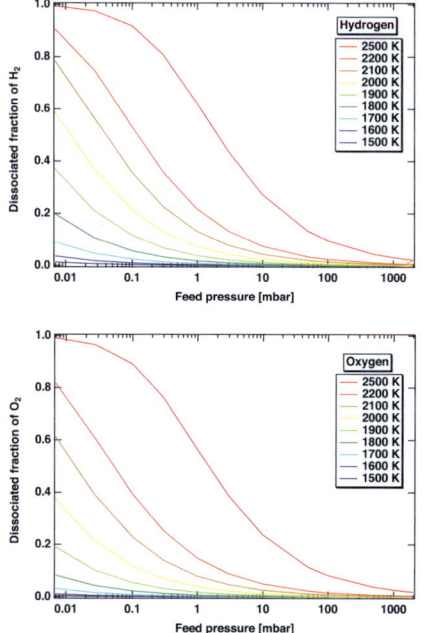

Figure 4.81: The **upper graph** shows the dissociated fraction of hydrogen atoms in dependence of the feeding pressure with molecular hydrogen for several temperatures of the capillary. The **lower graph** shows the dissociated fraction of oxygen atoms in dependence of the feeding pressure with molecular oxygen for several temperatures of the capillary.

4.9 Molecular reaction chemicurrent studies with MIS sensors

ies (using high pressure) delivered some ms for the life time of e.g. vibrationally excited molecular oxygen [395–397]. Therefore one could expect that these states live long enough to reach the MIS sensor after the excitation. Thus, I will have now some closer look on the possible excitations of vibrational states of the used gases.

In the first capillary measurements, the available excitation temperature was around 1700 K (compare section 4.9.2). The translational energy of such particles as O_2, H_2 or D_2 effusing out of the capillary would be $\approx$ 290 meV ($E = \frac{3}{2} \cdot k_B \cdot T + \frac{1}{2} \cdot k_B \cdot T = 2 \cdot k_B \cdot T$). Taking now possible vibrational excitations from the ground state into account, one can estimate roughly the fraction of particles $Vib_{frac}(v = 1)$ that may be vibrationally excited via the heated capillary, using a Boltzmann distribution.

molecule	ω_e [cm^{-1}, meV]	$\omega_e \cdot x_e$ [cm^{-1}, meV]	$\omega_e \cdot y_e$ [cm^{-1}, μeV]
O_2	$X^3\Sigma_g^-$ = 1580 , 196	12, 1	0.05, 7
D_2	$X^1\Sigma_g^+$ = 3118 , 386	64, 8	1.25, 155
H_2	$X^1\Sigma_g^+$ = 4395 , 515	117, 15	0.29, 36

Table 4.7: Vibrational quanta of the ground state and the products of the anharmonicity constants with the ground state quanta of O_2 [398], D_2 [399] and H_2 [399].

The vibrational quanta ω_e in the electronic ground state of oxygen, deuterium and hydrogen and the products of the anharmonicity constants (x_e, y_e) with the ground state quanta are given in table 4.7. With these values one can calculate the fraction Vib_{frac} of the number of the first excited state, $N(E_{v=1}(T))$, and the number of all states $\sum_{v=0}^{n} N(E_v(T))$ of the vibrationally excited molecules at a temperature $T = 1700\,\text{K}$ (of the heated capillary) with the Boltzmann constant k_B, the Planck constant h and the velocity of light c via equations 4.51, 4.52, 4.53 and 4.54 [400]. The obtained data are listed in table 4.8.

$$E_{(v)} = h \cdot c \cdot \omega_e \cdot (v + \frac{1}{2}) - h \cdot c \cdot \omega_e \cdot x_e \cdot (v + \frac{1}{2})^2 + h \cdot c \cdot \omega_e \cdot y_e \cdot (v + \frac{1}{2})^3 \quad (4.51)$$

4 Measurements, results and discussion

$$N(E_{v=1}(T)) = \sum_{v=1}^{1} e^{-\frac{E_{(v)}}{k_B \cdot T}} \tag{4.52}$$

$$\sum_{v=0}^{n} N(E_v(T)) = \sum_{v=0}^{n} e^{-\frac{E_{(v)}}{k_B \cdot T}} \tag{4.53}$$

$$\text{Vib}_{\text{frac}}(v=1) = \frac{N(E_{v=1}(T))}{\sum_{v=0}^{n} N(E_v(T))} \tag{4.54}$$

molecule	$N(E_{v=1}(T))$	$\sum_{v=0}^{n} N(E_v(T))$	$\text{Vib}_{\text{frac}}(v=1)$
O_2	0.138	0.704	0.196
D_2	0.019	0.289	0.067
H_2	0.005	0.165	0.029

Table 4.8: Calculated values for the number in the first excited state, $N(E_{v=1}(T))$, partition function $\sum_{v=0}^{n} N(E_v(T))$, and the fraction of the particles in the first excited state, $\text{Vib}_{\text{frac}}(v=1)$, that are excited with 1700 K, according to equations 4.51, 4.52, 4.53 and 4.54 [400].

From the data listed in table 4.8 one can see that more oxygen molecules are vibrationally excited when the capillary is heated. However, this could guide one to the conclusion, that the measured signal for an oxygen fed heated capillary should be much larger than for a hydrogen fed heated capillary, but that is not the case (compare figure 4.80). Therefore a description of the experiment shown in section 4.9.2 just with the pure amount of excited molecules at a capillary temperature of 1700 K is not appropriate.

Thus in the following the previously calculated dissociated fractions α_{hydrogen} and α_{oxygen} are used to obtain some information about how many molecules are dissociated, and how many molecules are in the first excited state to get more insight into the experiment described in section 4.9.3. The calculations (equations 4.55, 4.56) are exemplarily performed for hydrogen, the calculations for oxygen can be done analogously.

The pressure of the atomic hydrogen species p_H present at a certain temperature is the product of the feeding pressure p of the capillary and the dissociated fraction α_{hydrogen} of hydrogen:

$$p_H = p \cdot \alpha_{\text{hydrogen}} \tag{4.55}$$

These values of p_H are plotted together with the particle induced signal (compare figure 4.80), observed in the experiment (see section 4.9.3), in an Arrhenius plot as the natural logarithm versus the inverse temperature of the heated capillary (1 bar hydrogen gas feeding) in the upper graph of figure 4.82.

When one assumes that a fraction of the part of the hydrogen molecules that is not dissociated is vibrationally excited to the first state ($\nu = 1$), the pressure of the molecular hydrogen $p_{H_2}(\nu = 1)$ that is vibrationally excited in the first state can be derived by the following equation, using $Vib_{\text{frac}}(\nu = 1)$ from equation 4.54:

$$p_{H_2}(\nu = 1) = (p - p_H) \cdot Vib_{\text{frac}}(\nu = 1) \tag{4.56}$$

These values of $p_{H_2}(\nu = 1)$ are plotted together with the particle induced signal (compare figure 4.80), observed in the experiment (see section 4.9.3) in an Arrhenius plot as the natural logarithm versus the inverse temperature of the heated capillary (1 bar hydrogen gas feeding) in the lower graph of figure 4.83.

When these data in figure 4.82 are now compared, it seems to be that the experimental data can be interpreted in that way that just in the case where the highest temperatures ($460 \cdot 10^{-6} - 490 \cdot 10^{-6} \frac{1}{K}$) are used, atomic hydrogen is formed in a notable amount, what is in agreement with literature data [42, 104, 105]. In the medium temperature range ($500 \cdot 10^{-6} - 590 \cdot 10^{-6} \frac{1}{K}$) the experimental data observed can be best described assuming that the amount of molecular hydrogen in the vibrationally first excited state is large enough to explain the experiment. The deviations of the experimental data from both theoretical descriptions for the lowest used temperatures cannot yet be explained.

As mentioned, the same calculations using equations 4.55, 4.56 are performed for oxygen. The values for p_O are plotted together with the particle induced signal (compare figure 4.80), observed in the experiment (see section 4.9.3) in an Arrhenius plot as the natural logarithm versus the inverse temperature of the heated capillary (1 bar oxygen gas feeding) in the upper graph of figure 4.82.

The values of $p_{O_2}(\nu = 1)$ are plotted together with the particle induced signal (compare figure 4.80), observed in the experiment (see section 4.9.3) in an Arrhenius plot as the natural logarithm versus the inverse temperature of the heated capillary (1 bar hydrogen gas feeding) in the lower graph of figure 4.83. When

4 Measurements, results and discussion

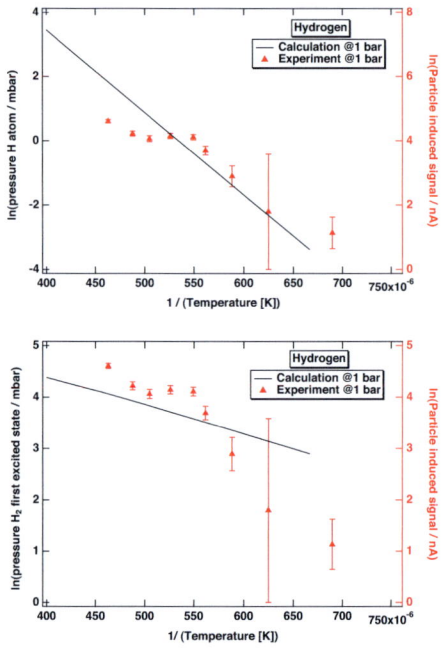

Figure 4.82: The **upper graph** shows a comparison of values, calculated via equation 4.55, of the natural logarithm of the pressure of atomic hydrogen and the experimentally observed particle induced signal (compare figure 4.80 in section 4.9.3) in dependence of the inverse temperature of the heated capillary. The **lower graph** shows the same experimental data in comparison to the natural logarithm of the hydrogen's first excited state pressure (calculated via equation 4.56) in dependence of the inverse temperature of the capillary. All data for feeding the heated capillary with 1 bar hydrogen gas.

4.9 Molecular reaction chemicurrent studies with MIS sensors

these data are now compared it seems to be that the experimental data cannot easily be described with the hypothesis that atomic oxygen or vibrationally excited oxygen is the reason for the observed signals. Maybe one can attribute the experimentally observed signals to molecular oxygen that is vibrationally excited in the first state in the temperature range between $550 \cdot 10^{-6} \frac{1}{K}$ and $590 \cdot 10^{-6} \frac{1}{K}$. For a more detailed explanation, further studies should be carried out.

However, I would conclude that I measured chemicurrent signals of atomic species and vibrationally excited molecules with the MIS sensor. Chemicurrent signals due to atomic species were already detected with MIM systems before [7–9, 42]. However, with the here used MIS systems even chemicurrents, that can be attributed to vibrationally excited molecules in the first state, are measurable. This opens a wide field for the detection of small electronic excitations, induced by incoming molecules. For example one could do this via an excitation of molecules with a laser. Therefore the effusive gas molecules or a molecular beam could be intersected with a a tunable laser, enabling state selective excitations. Then different specific excitations could be carried out and the signal due to excited molecules should be measurable, for example in experiments with excited molecular beams (excited by the optical pumping with the tunable laser). Furthermore, one could perform more experiments in the temperature regime where the atomic species is formed and study their signals in more detail. However, this goes beyond the scope of this work, but is an interesting perspective.

4.9.5 Time dependent trace of the chemicurrent signals

With the laser set-up, suggested in the previous section 4.9.4, it would as well be possible to determine the time dependent trace of the chemicurrent signals. The laser could then be used as well as a kind of shutter and thus excited or not excited molecules could be dosed towards the MIS chemicurrent sensor. However, first experiments regarding this trace can be carried using the heated capillary set-up with the shutter [8, 9], as used in section 4.9.2. The obtained curves are shown in figure 4.84. Here the recorded molecular chemicurrents normalized to the smallest probe flux of $1.8 \cdot 10^{16}$ particles/s at $t = 0$ s are shown for different fluxes. In the course of the experiment the mechanical shutter was opened twice (see indication in figure 4.84). As mentioned before, the difference in the height of the signal can be attributed to a photo induced signal when the difference between the opening and closing of the shutter is evaluated. The offset of this box like structure is due to the higher amount of excited particles, resulting in the higher observable currents as mentioned related to figure 4.78 before.

4 Measurements, results and discussion

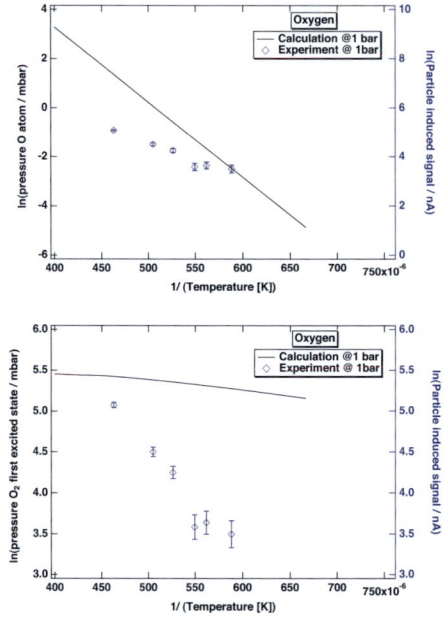

Figure 4.83: The **upper graph** shows a comparison of values, calculated via equation 4.55, of the natural logarithm of the pressure of atomic oxygen and the experimentally observed particle induced signal (compare figure 4.80 in section 4.9.3) in dependence of the inverse temperature of the heated capillary. The **lower graph** shows the same experimental data in comparison to the natural logarithm of the oxygen's first excited state pressure (calculated via equation 4.56) in dependence of the inverse temperature of the capillary. All data for feeding the heated capillary with 1 bar oxygen gas.

4.9 Molecular reaction chemicurrent studies with MIS sensors

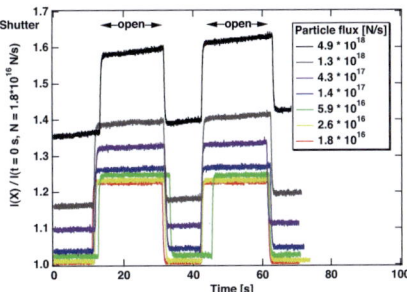

Figure 4.84: Chemicurrent of hydrogen excited by a capillary heated to 1700 K (as in figure 4.78) for different fluxes normalized to the signal at $t = 0\,\text{s}$ for a flux of $1.8 \cdot 10^{16}\frac{N}{s}$. Shutter opening periods of $\approx 20\,\text{s}$ are labeled. Results shown for +0.01 V bias voltage applied on the gold electrode.

For the highest particle fluxes used, a drift of the signal to higher current values occurs. This can be attributed to an insufficient pumping of the large amount of hydrogen flowing into the chamber, resulting in an increasing particle flux.

However, one can get valuable information out of these time dependent traces of the chemicurrent signal. To ease this, the signal for $5.9 \cdot 10^{16}$ particles per second is compared in figure 4.85 with a curve attributed to a pure photo current. This photo current curve is obtained when the heated capillary is operated with the same parameters as in the hydrogen experiment, just without any hydrogen feed.

One can clearly see that the photo induced signal shows a trace with rectangular shape. On the other hand the chemicurrent signals show an exponential slope, a bit comparable to the time dependent chemicurrent signals for atomic hydrogen in the literature [7–9]. When the shutter is opened it takes $\approx 1.3\,\text{s}$ until the current reaches its final value for the hydrogen induced signal, whereas the photo induced signal reaches its final value quite immediately (compare figure 4.85). When closing the shutter the hydrogen induced signal reaches its static value after $\approx 2.1\,\text{s}$. Again the photo induced signal reaches its starting value quite immediately. One must note that the reason for these time constants is not the flight time between the position of the shutter and the MIS sensor. However, the underlying chemical process cannot yet be determined.

4 Measurements, results and discussion

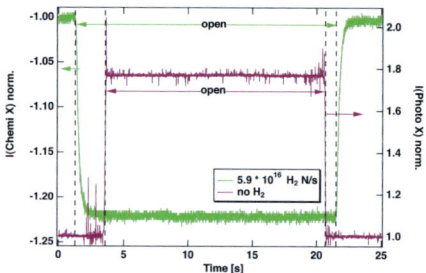

Figure 4.85: Comparison of the normalized chemicurrent of hydrogen excited by a capillary heated to 1700 K (as in figure 4.78) and a pure photo excitation without any hydrogen feed but a heated capillary (1700 K). Shutter opening periods are labeled. Results shown for +0.01 V bias voltage applied on the gold electrode.

4.9 Molecular reaction chemicurrent studies with MIS sensors

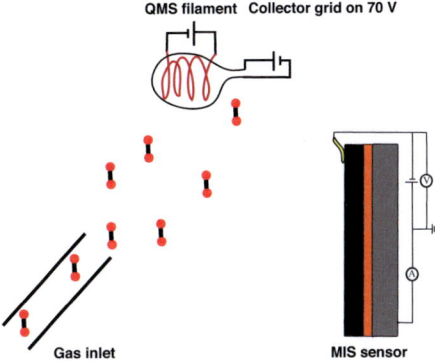

Figure 4.86: Schematic drawing of the set-up used for an indirect dosage of excited species on a MIS sensor. The gas inlet, the QMS with the heated filament and the collector grid are not in line with the MIS sensor.

4.9.6 Indirect dosage of excited species on a MIS sensor

Significant chemicurrents can be measured with platinum MIS sensors, even when only the QMS is operating in the UHV chamber and molecular gases as oxygen, deuterium or hydrogen are effusively dosed into the chamber. Such chemicurrents can be observed as well when just the QMS filament and the collector grid (on 70 V) are operated without any high voltage. Therefore one has to be aware of such sources of errors where particles can be excited under not defined conditions, as just via an operating QMS filament, when chemicurrent experiments are carried out. However, one can describe the sensitivity of the platinum MIS in such an experiment, which is sketched in figure 4.86. Furthermore I would like to emphasize that the gas inlet, the operating QMS and the MIS sensor are not set in a line. Several collisions of the particles between each other or with the walls have to occur until they reach the sensor.

Figure 4.87 shows the chemicurrents for different gases. In the upper graph of figure 4.87 the chemicurrent is plotted versus the pressure in the chamber and the corresponding particle fluxes. The recorded chemicurrent signals show different signals for the different gases, furthermore a kind of saturation value ($\approx -225\,\text{nA}$) for higher particle fluxes can be observed in the case of oxygen. This negative polarity denotes again a net electron current from the bottom silicon to the top platinum layer, as mentioned before. To enable a better view on the chemicurrents of the other gases, a zoom for smaller chemicurrents is pre-

179

4 Measurements, results and discussion

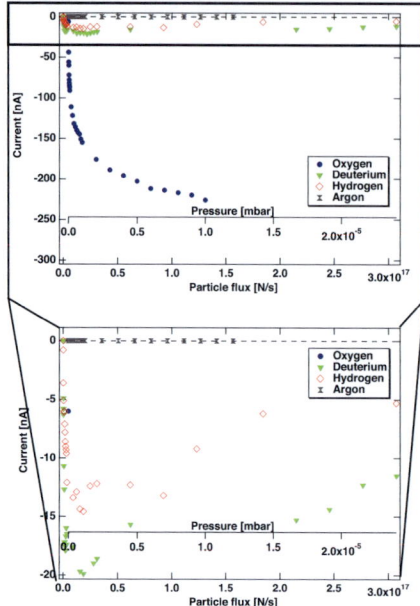

Figure 4.87: Chemicurrent for different gases plotted versus the gas pressure and the corresponding particle fluxes. The **upper graph** shows an overview and the **lower graph** shows a zoom for smaller chemicurrents.

sented in the lower graph of figure 4.87. A maximum in the chemicurrent signal is observed for deuterium (-19.9 nA) and hydrogen (-14.5 nA) for a dosage of $\approx 10^{-6}$ mbar, respectively $\approx 10^{16}$ particles per second.

Even a better view on the difference in the chemicurrent signals of the probed gases can be obtained, when the signals are plotted in terms of a yield. This yield can be obtained via the simple assumption that every incident particle induces a chemicurrent signal (what in reality will be an overestimation since it is very probable that not every gas particle can be excited by the QMS before it reaches the MIS sensor). These values are drawn in the upper graph of figure 4.88 versus the particle flux. When the yield is plotted versus the logarithm of the particle flux (see lower graph of figure 4.88) a maximum in the yield can be observed for oxygen, deuterium and hydrogen. From this one can put the signals of the molecules effusing out of the gas inlet system at ambient temperatures, which are excited via the QMS, in the following order (values taken at the maximum of

4.9 Molecular reaction chemicurrent studies with MIS sensors

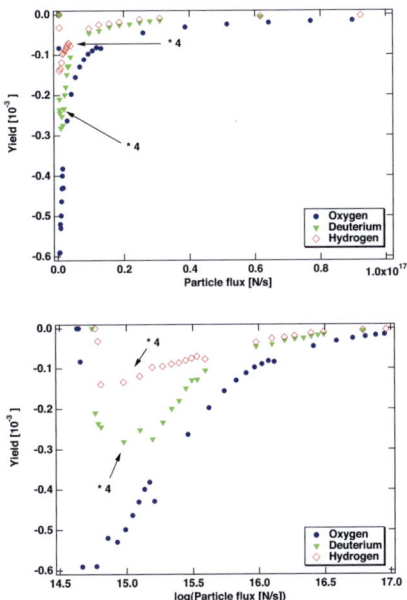

Figure 4.88: Chemicurrent yield plotted versus the particle flux (**upper graph**) and the chemicurrent yield plotted versus the logarithm of the particle flux (**lower graph**) for different gases. Yields for hydrogen and deuterium are multiplied with a factor of 4.

the yield, compare figure 4.88 and table 4.9):

$$\text{Oxygen} > \text{Deuterium} > \text{Hydrogen} \qquad (4.57)$$

dosed gas	pressure [mbar]	particle flux [N/s]	exp. chemicurrent [nA]	exp. chemo yield
O_2	$2 \cdot 10^{-8}$	$5.9 \cdot 10^{14}$	-44.0	$-5.9 \cdot 10^{-4}$
D_2	$2 \cdot 10^{-8}$	$9.5 \cdot 10^{14}$	-10.7	$-7.0 \cdot 10^{-5}$
H_2	$1 \cdot 10^{-8}$	$6.4 \cdot 10^{14}$	-3.6	$-3.5 \cdot 10^{-5}$

Table 4.9: Experimentally observed chemicurrents and chemoyields for oxygen, deuterium and hydrogen species at denoted pressures, respectively particle fluxes, evaluated at the maximum of the yield. Data taken from figure 4.88.

However, the monoatomic gas argon does not give rise to any signal. This chemical selectivity is underpinned when the same experiment is repeated. Then the subsequent sensor currents (for the molecular gases) are smaller than for the first experiment. However, when a cleaning via a sputtering process (with Ar^+ ions of e.g. 90 eV) is carried out in between, then the same results are observed as if one would start with a clean surface.

When one assumes that the QMS filament is around 1700 K hot (The filament shows approximately the same glowing as the capillary used in section 4.9.2), one can compare the observed chemicurrent signals with the fraction of the particles in the first excited state $Vib_{frac}(v = 1)$ that are excited at 1700 K (compare table 4.8). Interestingly the observed order for the chemicurrent signals is in coincidence with the order of the fraction of the particles in the first excited state $Vib_{frac}(v = 1)$. In the experiment, the chemicurrent for oxygen is around 4 times larger than the one for deuterium, and around 12 times larger than for hydrogen (compare table 4.9). The fraction of the particles in the first excited state for oxygen is around 4 times larger than for deuterium, and around 14 times larger than for hydrogen. So this fits quite well. Probably the reason for this observed order is the excitation process at the QMS filament. Meaning more excited oxygen than deuterium or hydrogen species are present after an excitation process at the QMS filament. A first clue for further studies regarding this process at the QMS filament can be obtained when the data of figure 4.88 are plotted as a mass normalized yield versus the logarithm of flux (see figure 4.89). Then nearly the

4.9 Molecular reaction chemicurrent studies with MIS sensors

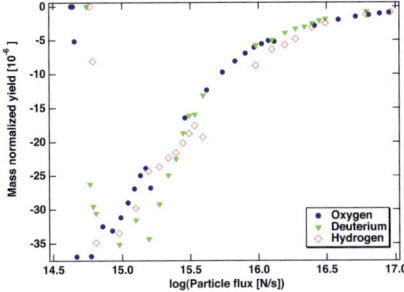

Figure 4.89: Mass normalized chemicurrent yield plotted versus the logarithm of the particle flux for oxygen, deuterium and hydrogen.

same mass normalized yields are found for oxygen, deuterium and hydrogen.

Additionally the previously mentioned maximum in the yield (compare figure 4.88 and table 4.9) can be compared with a calculated yield when a MIM detector would be used. Therefore the maximum yield for oxygen of $5.9 \cdot 10^{-4}$ is taken as a comparative value. In a previous section (4.8) I compared the photosensitivity of MIM and MIS systems (see also figure 4.65). The ratio of the sensitivities was smaller than 10^{-5} for an excitation energy of 2 eV when the MIS sensor is compared with a tantalum based MIM. Hence the experiment shown for the MIS system in figures 4.87, 4.88 would give a maximum yield of $\approx 10^{-9}$, if it would be accomplished with a tantalum based MIM system. This corresponds to a maximum current of $\approx 5\,\text{pA}$, which is on the order of the noise level of the system used. Surely this is the reason why MIMs did not deliver a molecule induced chemicurrent signal up to now.

4.9.7 Effusion of deuterium and oxygen mixtures

To study the water formation reaction one has to choose the right temperature range for the water formation reaction on the e.g. polycrystalline platinum top surface of the MIS sensor (compare section 2.4). It would be interesting to carry out the water formation reaction in the temperature range of 170 − 220 K [98, 100] for which water desorption takes place, as well as below and above this temperature range. Then the previously mentioned studies (see section 2.4) of the water formation reaction in the literature [93–101] can be put together with the new chemicurrent experiments to deliver more insights into the dissipation processes during the water formation reaction.

To perform some first experiments at ambient temperatures an effusive source for deuterium and oxygen mixtures is used. In these experiments a Si–SiOx(1 nm / 4 nm)–Pt sensor is exposed (on the 1 nm thick oxide area) at ambient temperatures to different mixtures of deuterium and oxygen for several minutes (QMS is operating). In addition to the MIS sensor signals, the signal from the QMS in the chamber is recorded. However, in contrast to the reactants, no increase of the mass 20 signal indicating the formation of water out of the deuterium and oxygen supply was detected.

As expected one can observe chemicurrents with the sensor, as demonstrated in figure 4.90. However, it is quite surprising that one does not observe much larger currents than presented in figures 4.87, 4.88 in section 4.9.6. Furthermore, there is a difference in the magnitude of the signals (especially in the case of oxygen) when the experiments are carried out with a 'freshly prepared and cleaned' MIS sensor, as it was done in section 4.9.6, or here where several mixtures of deuterium and oxygen are probed. To have a closer look on the chemicurrents from the gas mixtures, the measured chemicurrents are plotted as usual versus the pressure and the particle flux for different mixtures of deuterium and oxygen (upper graph of figure 4.90). One can see similar to the experiments presented before the increase and saturation of the chemicurrent for larger fluxes. In the middle graph of figure 4.90 one can identify the previously mentioned maxima in the yield around 10^{15} particles/s for all mixtures.

The trend for different mixtures of deuterium and oxygen can be better seen in the lower graph where the yields for selected particle fluxes of the gas mixtures are plotted. From a simple point of view one would expect a larger yield (indicating an excess energy of the water formation reaction), e.g. at a mixture of $0.66\,D_2/0.33\,O_2$ to form with two deuterium and one oxygen molecule two water molecules, than for pure molecular oxygen. But the data does not show this.

Rather it seems to be the case that the amount of oxygen (leading to larger signals, compare figures 4.87, 4.88) plays the key role and the deuterium acts more like a dilutor. More experiments are necessary to study this in more detail. It would be ideal when also the water formation could be quantified with the QMS. Therefore, one should carry out this experiment at various temperatures. Then one should be able to study the water formation reaction on the MIS top surface in terms of chemicurrent studies.

4 Measurements, results and discussion

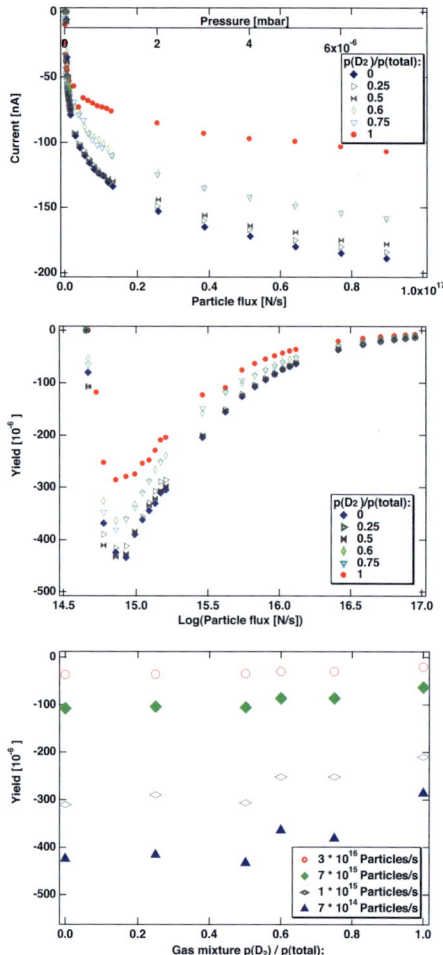

Figure 4.90: **Upper graph:** Molecular chemicurrent signal in dependence of the gas pressure respectively the particle flux introduced into the apparatus for different mixtures of deuterium and oxygen. **Middle graph:** Molecular chemicurrent yield in dependence of the logarithm of the particle flux. **Lower graph** shows the molecular chemicurrent yield for selected particle fluxes in dependence of the deuterium and oxygen ratio.

5 Summary and Outlook

It was the aim of this work to study electronic dissipation processes during chemical reactions on surfaces. In particular an aim was to study these processes with the help of chemicurrent sensors, molecular beams and effusive gas sources in terms of adsorption of molecular species (e.g. on platinum top electrode of a chemicurent sensor). Furthermore, chemical reactions of species accommodated on the surface with species impinging from a molecular beam, should be studied.

I was able to show, that with the formerly existing sensors (aluminum oxide and tantalum oxide based) no analysis of molecular reactions in terms of electronic dissipation processes are possible. Hence, I characterized the sensors regarding their backelectrode, insulating oxide, top electrode, insulator thickness, metal thickness and internal photoemission in order to obtain more information about what could be possibly improved.

Then a series of different sensors starting from the established aluminum oxide and tantalum oxide based sensors to new titanium oxide and new silicon oxide based sensors was characterized. This series of sensors ranges from high internal barriers (3.0 eV for aluminum oxide) to low internal barriers ($<$ 1 eV for silicon oxide based sensors). Out of these results I presented in this thesis those for titanium oxide based sensors. These titanium oxide based sensors exhibit a two oders of magnitude larger photosensitivity than tantalum oxide based sensors (see figure 5.1). However, these titanium oxide based sensors lack in their stability, in e.g. atomic hydrogen chemicurrent experiments their dielectric properties change during the experiments. These sensors have to be improved in the future by adding a surfactant on the titanium oxide layer to enable a layer by layer growth of the top electrode.

Following a different route, namely switching from metal–insulator–metal to metal–insulator–semiconductor sensors, I set up a new line of silicon based sensors with an electrochemically prepared oxide of desired thickness. These MIS sensors can even be manufactured with stepped oxide thicknesses what enables a study of devices with different thick oxide terraces similar to an sensor array. These sensors unite the advantage of a very high sensitivity, which is more than

5 Summary and Outlook

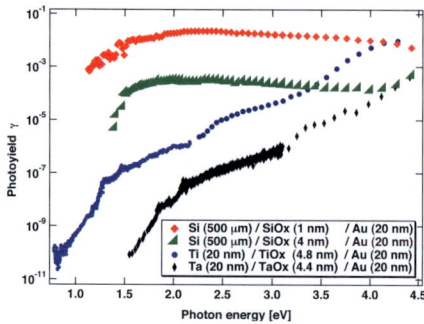

Figure 5.1: Comparison of the photoyields of the main working horse chemicurrent sensors used in this work.

five orders of magitude larger in internal photoemission studies than for comparable tantalum oxide based sensors (compare figure 5.1), with the advantage of high stability (no change of the dielectric properties was observed in chemicurrent experiments). First characterization experiments were carried out using MIS sensors with a 20 nm gold top electrode or a 7 nm thin platinum top electrode. This high sensitivity can be seen as well in experiments with low energy (E_{kin} = 90 eV) Ar^+ or H_2^+ ions. Here a current of some ten nA, respectively a yield of some ten percent, is observed.

Finally, I studied species with kinetic energies in the range of the chemisorption enthalpy, which is typically in the range of 0.5 to 2.0 eV. For these experiments I returned to the fundamental works of Irving Langmuir [16–19]. I carried out experiments regarding the chemical selectivity of the detectors, using a stream of excited hydrogen molecules and hydrogen atoms. Here the excitation and radical formation was achieved by the interaction of ground state molecules with a hot tungsten surface, according to the pioneering experiments of Langmuir. I was able to measure chemicurrent signals due to atomic species and vibrationally excited molecules with the MIS sensor. Chemicurrent signals due to atomic species were already measured with MIM systems before [7–9, 42]. However, with the here used MIS systems even chemicurrents, that can be attributed to vibrationally excited molecules in the first state, were measurable for the first time.

Additionally, I was able to attribute the conduction mechanism in these very sensitive MIS sensors as an excited hole conduction process based on the study of the internal photoemission, low energy ion experiments and the chemicurrent studies of vibrationally excited molecules.

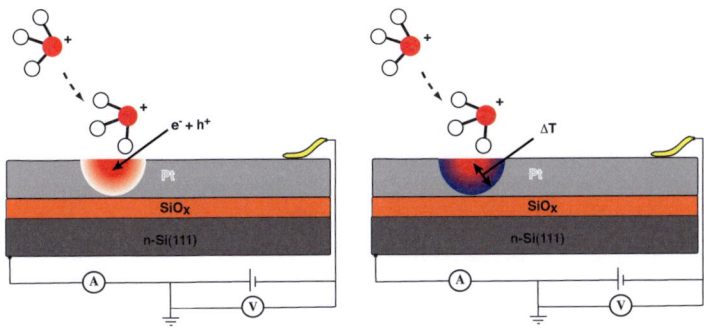

Figure 5.2: Schematic drawing for a comparison of the processes that induce a sensor current in terms of electronic excitations (electron–hole pair generation) that underlie the non-adiabatic processes (left side), and in terms of the small local heating (right side) when a H_3O^+ molecule approaches a platinum-MIS.

Furthermore, studies regarding the cleaning of the sensor's surface, studies with the molecular beam and temperature programmed desorption are carried out for the first time on top of a chemoelectronic sensor - all without changing the dielectric properties of the sensors.

Concluding, I set up and characterized a new kind of low barrier sensor that has the potential to be used for the study of non-adiabatic processes.

Additionally, to the deviations from the electronic equilibrium, deviations from thermal equilibrium with very small temperature changes can now be monitored with a reasonable efficiency, as we showed in a recent publication [379]. In this work we showed a model calculation for a metal which is moderately heated ($\Delta T = 10^{-4}$ K) and separated via an e.g. 1 eV high barrier for electrons from a second metal, respectively a semiconductor layer, the temperature of which is kept constant. This temperature difference across the device leads to an electron current from the warmer to the colder electrode. In the model calculation we found a current density value of $10\,\text{nA} \cdot \text{cm}^{-2}$ for 1 eV barrier height and the mentioned temperature difference.

The detection of smallest temperature differences can now be used as a further diagnostic tool for the study of charge transfer reactions on surfaces. In a gedankenexperiment an H_3O^+ molecule approaches a metal surface (e.g. platinum top electrode of a MIS sensor, compare figure 5.2). This induces then the previously mentioned electronic excitations (sketched on the left side of figure

5 Summary and Outlook

5.2) as well as a small localized heating (sketched on the right side of figure 5.2). In the latter process, the metal surface is assumed to be locally heated by some mK, on a nm sized area, when H_3O^+ molecules are approaching. Since the heat transport in metals is dominated by electrons, the heat will come to the inner sensor interface and leads to a sensor current within some 10 fs. This process can well compete with the heat transport to the water layer, since this transport is reliant on vibrations. So, the sensor current which is based on the transport of electrons over the internal barrier can be seen as a measure for the H_3O^+ induced local heating of the top electrode. This local heating would also promote the electron transfer from the metal to the H_3O^+. If one would for example detect a reaction induced current over the internal barrier of x eV height, then one could state that the charge transfer reaction from the metal to the H_3O^+ is at least also driven by charge carriers with x eV excess energy.

Schmickler and coworkers showed recently that the activation barrier for hydrogen adsorption at the metal–liquid interface is $0.5 - 0.7$ eV for silver–acid interfaces and gold–acid interfaces [401]. This barrier height is comparable to the barrier height in silicon–silicon oxide–gold systems which I presented in my work. If the transport over this barrier height on for example platinum surfaces is mainly induced by an adsorbate induced heating of the platinum substrate, one would detect a sensor current of approximately the same magnitude as the charge transfer current from the platinum to the H_3O^+ species. If one wants to do an energy selective detection of the charge carriers, one of the other sensor systems which I presented in my work, can be chosen. For 1 eV high barriers for example titanium oxide based sensors would be appropriate, for 1.5 eV tantalum oxide sensors and for 3 eV high activation barriers one can choose aluminum oxide based sensor systems.

However, with the low barrier MIS systems a separation of the signal into non-adiabatic processes and smallest temperature differences is not possible. But a separation will be possible when the other systems presented in this work with higher internal barrier are used. Going from MIS to titanium oxide via tantalum oxide to aluminum oxide based MIM sensors, the signals related to the small temperature differences will vanish more and more, whereas some non-adiabatic processes may still be possibly detected.

Consequently, one can also think about letting the barrier (the oxide system) unchanged and changing the top metals. If for example the discharging reaction of H_3O^+ would lead to a measurable sensor current on platinum, but for example a weaker signal on gold, one can denote that the platinum electrode's thermodynamic state show a higher deviation from thermodynamic or electronic equilibrium. Always keeping in mind, that gold shows a hydrogen overpotential and platinum does not. Thus, the studies on deviations from thermodynamic and electronic equilibrium will lead us then to the development of microscopic models for overpotentials.

The concept of monitoring deviations from thermal and electronic equilibrium is expandable to other surface reactions due the wide choice of sensor systems. It should be mentioned, that no other technique is presently able to study these small effects. For a surface chemist, up to now these effects were on the dark side of the moon. With the technique developed in this work it will be possible to shed some light on the 'dark side' of several surface reactions.

Bibliography

[1] G. Ertl, *Nobel lecture* (Nobel Prize Chemistry, Stockholm, 2007).

[2] G. Ertl, in *Reactions at solid surfaces* (John Wiley & Sons, Hoboken, New Jersey, 2009), p. 103.

[3] B. Gergen, H. Nienhaus, W. H. Weinberg, and E. W. McFarland, Science **294**, 2521 (2001).

[4] H. Nienhaus, H. S. Bergh, B. Gergen, A. Majumdar, W. H. Weinberg, and E. W. McFarland, Applied Physics Letters **74**, 4046 (1999).

[5] D. J. Auerbach, Science **294**, 2488 (2001).

[6] H. Nienhaus, B. Gergen, H. S. Bergh, A. Majumdar, W. H. Weinberg, and E. W. McFarland, Phys. Rev. Lett. **82**, 446 (1999).

[7] B. Mildner, E. Hasselbrink, and D. Diesing, Chemical Physics Letters **432**, 133 (2006).

[8] B. Schindler, Ph.D. thesis, Fachbereich Chemie, Universität Duisburg-Essen, 2009.

[9] B. Schindler, D. Diesing, and E. Hasselbrink, The Journal of Chemical Physics **134**, 034705 (2011).

[10] E. Hasselbrink, Current Opinion in Solid State and Materials Science **10**, 192 (2006).

[11] E. Hasselbrink, Surface Science **603**, 1564 (2009).

[12] E. G. Karpov and I. Nedrygailov, Applied Physics Letters **94**, 214101 (2009).

[13] E. G. Karpov and I. Nedrygailov, Phys. Rev. B **81**, 205443 (2010).

[14] D. A. Kovacs, J. Winter, S. Meyer, A. Wucher, and D. Diesing, Physical Review B (Condensed Matter and Materials Physics) **76**, 235408 (2007).

[15] D. Kovacs, A. Golczewski, G. Kowarik, F. Aumayr, and D. Diesing, Phys. Rev. B **81**, 075411 (2010).

[16] I. Langmuir, Journal of the American Chemical Society **34**, 1310 (1912).

[17] I. Langmuir, Journal of the American Chemical Society **34**, 860 (1912).

[18] I. Langmuir, Journal of the American Chemical Society **37**, 417 (1915).

[19] I. Langmuir, Journal of the American Chemical Society **38**, 1145 (1916).

[20] T. Greber, Surface Science Reports **28**, 1 (1997).

[21] B. Kasemo, Physical Review Letters **32**, 1114 (1974).

[22] D. Andersson, B. Kasemo, and L. Wallden, Chemical Physics Letters **111**, 593 (1984).

[23] A. Böttcher, R. Imbeck, A. Morgante, and G. Ertl, Physical Review Letters **65**, 2035 (1990).

[24] L. Hellberg, J. Strömquist, B. Kasemo, and B. I. Lundqvist, Physical Review Letters **74**, 4742 (1995).

[25] A. Böttcher, R. Grobecker, T. Greber, and G. Ertl, Chemical Physics Letters **208**, 404 (1993).

[26] T. Greber, R. Grobecker, A. Morgante, A. Böttcher, and G. Ertl, Physical Review Letters **70**, 1331 (1993).

[27] T. Greber, K. Freihube, R. Grobecker, A. Böttcher, K. Hermann, G. Ertl, and D. Fick, Physical Review B **50**, 8755 (1994).

[28] J. Robertson, Reports on Progress in Physics **69**, 327 (2006).

[29] R. Sessoli, H. L. Tsai, A. R. Schake, S. Wang, J. B. Vincent, K. Folting, D. Gatteschi, G. Christou, and D. N. Hendrickson, Journal of the American Chemical Society **115**, 1804 (1993).

[30] C. A. Baumann, R. J. Van Zee, and W. Weltner, The Journal of Physical Chemistry **86**, 5084 (1982).

[31] M. Born and R. Oppenheimer, Annalen der Physik **389**, 457 (1927).

[32] A. M. Wodtke, Y. Huang, and D. J. Auerbach, The Journal of Chemical Physics **118**, 8033 (2003).

[33] J. D. White, J. Chen, D. Matsiev, D. J. Auerbach, and A. M. Wodtke, Nature **433**, 503 (2005).

[34] P. W. Anderson, Physical Review Letters **18**, 1049 (1967).

[35] E. Müller-Hartmann, T. V. Ramakrishnan, and G. Toulouse, Phys. Rev. B **3**, 1102 (1971).

[36] E. Müller-Hartmann, T. V. Ramakrishna, and G. Toulouse, Solid State Communications **9**, 99 (1971).

[37] H. Nienhaus, Surf. Sci. Rep. **45**, 1 (2002).

[38] G. R. Darling and S. Holloway, Reports on Progress in Physics **58**, 1595 (1995).

[39] M. Scheele, Diploma thesis, Fakultät ür Chemie, Universität Duisburg-Essen, 2010.

[40] H. Nienhaus, private communications (2010).

[41] A. D. Smith, M. Tinkham, and W. J. Skocpol, Physical Review B **22**, 4346 (1980).

[42] B. Mildner, Diploma thesis, Fachbereich Chemie, Universität Duisburg-Essen, 2005.

[43] S. Glass and H. Nienhaus, Physical Review Letters **93**, 168302 (2004).

[44] X. Liu, B. R. Cuenya, and E. W. McFarland, Sensors and Actuators B: Chemical **99**, 556 (2004).

[45] E. Hasselbrink, Surface Science **603**, 1564 (2009).

[46] D. Diesing and H. Nienhaus, Essener Unikate **32**, 52 (2007).

[47] S. Sze, *Physics of semiconductor devices* (Wiley Interscience, New York Chichester Brisbane Toronto Singapore, 1981).

[48] O. Marti and A. Plettl, Vorlesungsskript Physikalische Elektronik und Messtechnik Universität Ulm, 2007.

[49] G. Scoles, *Atomic and Molecular Beam Methods Vol. I* (Oxford Universitty Press, New York, Oxford, 1988).

Bibliography

[50] H. Pauly, *Atom, Molecule, and Cluster Beams II* (Springer Verlag, Berlin, Heidelberg, 2000).

[51] H. Pauly, *Atom, Molecule, and Cluster Beams I* (Springer Verlag, Berlin, Heidelberg, 2000).

[52] C. T. Campbell, G. Ertl, H. Kuipers, and J. Segner, The Journal of Chemical Physics **73**, 5862 (1980).

[53] C. T. Campbell, G. Ertl, and J. Segner, Surface Science **115**, 309 (1982).

[54] C. T. Campbell, G. Ertl, H. Kuipers, and J. Segner, Surface Science **107**, 220 (1981).

[55] A. E. Wiskerke, F. H. Geuzebroek, A. W. Kleyn, and B. E. Hayden, Surface Science **272**, 256 (1992).

[56] P. D. Nolan, B. R. Lutz, P. L. Tanaka, J. E. Davis, and C. B. Mullins, The Journal of Chemical Physics **111**, 3696 (1999).

[57] M. Balooch, M. J. Cardillo, D. R. Miller, and R. E. Stickney, Surface Science **46**, 358 (1974).

[58] J. A. Barker and D. J. Auerbach, Surface Science Reports **4**, 1 (1984).

[59] T. H. Lin and G. A. Somorjai, Surface Science **107**, 573 (1981).

[60] P. M. Holmblad, J. Wambach, and I. Chorkendorff, The Journal of Chemical Physics **102**, 8255 (1995).

[61] C. T. Campbell, G. Ertl, H. Kuipers, and J. Segner, Surface Science **107**, 207 (1981).

[62] J. Libuda and H. J. Freund, Surface Science Reports **57**, 157 (2005).

[63] R. Horn, Ph.D. thesis, Dissertation der Fakultät II - Mathematik und Naturwissenschaften der Technischen Universität Berlin, 2003.

[64] N. F. Ramsey, *Molecular Beams* (Oxford University Press, Oxford, 1990).

[65] A. E. Zarvin and R. G. Sharafutdinov, Journal of Applied Mechanics and Technical Physics **20**, 744 (1979).

[66] M. J. Pilling and P. Seakins, *Reaction Kinetics* (Oxford University Press, Oxford, 1995), pp. 100–106.

[67] O. Autzen, Ph.D. thesis, Fachbereich Chemie Universität Duisburg-Essen, 2005.

[68] J. G. Segner, Ph.D. thesis, Inaugural-Dissertation Fachbereich Chemie und Pharmazie Ludwig-Maximilians-Universität München, 1982.

[69] C. G. Eisenhardt, Ph.D. thesis, Inaugural Dissertation Fachbereich Biologie, Chemie, Pharmazie der Freien Universität Berlin, 2000.

[70] M. Binetti, Ph.D. thesis, Fachbereich Chemie Universität Essen, 2001.

[71] K. Lass, Diploma thesis, Fachbereich Chemie, Universität Duisburg-Essen, 2001.

[72] I. Meusel, Ph.D. thesis, Dissertation der Fakultät II - Mathematik und Naturwissenschaften der Technischen Universität Berlin, 2002.

[73] C. R. Arumainayagam and R. J. Madix, Progress in Surface Science **38**, 1 (1991).

[74] B. Riedmüller, F. Giskes, D. G. van Loon, P. Lassing, and A. W. Kleyn, Meas. Sci. Technol. **13**, 141 (2002).

[75] G. I. Dimov, in *Atom, Molecule, and Cluster Beams I by H. Pauly* (Springer Verlag, Berlin, Heidelberg, 2000).

[76] W. R. Gentry, in *Atomic Molecular Beam Methods, Vol. 1, ed. by G. Scoles* (Oxford University Press, Oxford, 1988), p. 54.

[77] W. R. Gentry and C. F. Giese, Review of Scientific Instruments **49**, 595 (1978).

[78] H. W. Lülf and P. Andresen, *Rarefied Gas Dynamics 13, Vol2. ed. by O.M. Belotserkovskii, M.N. Kogan, S.S. Kutateladse, and A.K. Rebrov* (Plenum Press New York, New York, 1985), p. 911.

[79] C. H. Hamann and W. Vielstich, *Elektrochemie* (WILEY-VCH, Weinheim, 2005).

[80] D. Hibbert, *Introduction to electrochemistry* (The Macmillan Press LTD, London, 1993).

[81] K. Stella, D. A. Kovacs, and D. Diesing, Electrochemical and Solid-State Letters **12**, H453 (2009).

[82] K. Stella, D. Bürstel, S. Franzka, O. Posth, and D. Diesing, Journal of Physics D: Applied Physics **42**, 135417 (2009).

[83] K. Stella, D. A. Kovacs, D. Diesing, W. Brezna, and J. Smoliner, Journal of The Electrochemical Society **158**, P65 (2011).

[84] K. Stella and D. Diesing, Journal of The Electrochemical Society **154**, C663 (2007).

[85] Y. Jeliazova, M. Kayser, B. Mildner, A.W. Hassel, and D. Diesing, Thin Solid Films **500**, 330 (2006).

[86] A.W. Hassel and M.M. Lohrengel, Electrochim. Acta **42**, 3327 (1997).

[87] A.W. Hassel and M.M. Lohrengel, Electrochim. Acta **40**, 433 (1995).

[88] J.W. Schultze and A.W. Hassel, in *Encyclopedia of Electrochemistry* (Wiley-VCH, Weinheim, 2003), Vol. 4, section 3.2, pp. 216–270.

[89] D. Bürstel, Diploma thesis, Fachbereich Chemie, Universität Duisburg-Essen, 2009.

[90] X. G. Zhang, in *Electrochemistry of silicon and its oxide* (Kluwer Academic Publishers, Plenum Press, New York, Boston, Dordrecht, London, Moscow, 2001), p. 103.

[91] J. Doebereiner, *Über neu entdeckte höchst merkwürdige Eigenschaften des Platins und die pneumatisch-capillare Thätigkeit gesprungener Gläser* (Schmid, Jena, 1823).

[92] J. Doebereiner, in *Die neuesten und wichtigsten physikalisch-chemischen Entdeckungen* (Schmid, Jena, 1823), p. 145.

[93] S. Völkening, K. Bedürftig, K. Jacobi, J. Wintterlin, and G. Ertl, Phys. Rev. Lett. **83**, 2672 (1999).

[94] B. Hellsing, B. Kasemo, S. Ljungstrom, A. Rosén, and T. Wahnstrom, Surface Science **189-190**, 851 (1987).

[95] E. Fridell, A.-P. Elg, A. Rosén, and B. Kasemo, The Journal of Chemical Physics **102**, 5827 (1995).

[96] F. Eisert, F. Gudmundson, and A. Rosén, Appl. Phys. B **68**, 579 (1999).

[97] K. Bedürftig, S. Völkening, Y. Wang, J. Wintterlin, K. Jacobi, and G. Ertl, The Journal of Chemical Physics **111**, 11147 (1999).

[98] C. Sachs, M. Hildebrand, S. Volkening, J. Wintterlin, and G. Ertl, The Journal of Chemical Physics **116**, 5759 (2002).

[99] I. K. Verheij and M. B. Hugenschmidt, Surface Science **416**, 37 (1998).

[100] L. K. Verheij, Surface Science **371**, 100 (1997).

[101] L. K. Verheij and M. B. Hugenschmidt, Surface Science **324**, 185 (1995).

[102] L. K. Verheij, M. Freitag, M. B. Hugenschmidt, I. Kempf, B. Poelsema, and G. Comsa, Surface Science **272**, 276 (1992).

[103] K. M. Ogle and J. M. White, Surface Science **139**, 43 (1984).

[104] K. Tschersich, J. Appl. Phys. **87**, 2565 (2000).

[105] K. Tschersich and V. von Bonin, J. Appl. Phys. **84**, 4065 (1998).

[106] K. Stella, Diploma thesis, Fachbereich Chemie, Universität Duisburg-Essen, 2008.

[107] D. A. Kovacs, T. Peters, C. Haake, M. Schleberger, A. Wucher, A. Golczewski, F. Aumayr, and D. Diesing, Phys. Rev. B **77**, 245432 (2008).

[108] D. Kovacs, T. Babkina, T. Gans, U. Czarnetzki, and D. Diesing, Journal of Physics D: Applied Physics **39**, 5224 (2006).

[109] J. Lindhard and M. Scharff, Phys. Rev. **124**, 128 (1961).

[110] H. D. Hagstrum, Phys. Rev. **96**, 336 (1954).

[111] P. Apell, Journal of Physics B: Atomic, Molecular and Optical Physics **21**, 2665 (1988).

[112] M. Suchanska, Progress in Surface Science **54**, 165 (1997).

[113] J. J. Cuomo, S. M. Rossnagel, and H. R. Kaufman, in *Handbook of Ion Beam Processing Technology - Principles, Deposition, Film Modification and Synthesis* (Noyes Publications, New Jersey, 1989), p. 2.

[114] R. Hellborg, H. J. Whitlow, and Y. Zhang, in *Ion Beams in Nanoscience and Technology* (Springer, Heidelberg Dordrecht London New York, 2009), p. 69.

[115] J. P. Biersack, Nuclear Instruments and Methods **182-183**, 199 (1981).

[116] J. P. Biersack and L. G. Haggmark, Nuclear Instruments and Methods **174**, 257 (1980).

[117] SRIM Ion Impact Software, http://www.srim.org, 2010.

[118] B. Gergen, H. Nienhaus, W. Weinberg, and E. McFarland, Science **294**, 2521 (2001).

[119] H. Nienhaus, H. S. Bergh, B. Gergen, A. Majumdar, W. H. Weinberg, and E. McFarland, J. Vac. Sci. Technol. A **17**, 1683 (1999).

[120] H. Nienhaus, H. Bergh, B. Gergen, A. Majumdar, W. H. Weinberg, and E. W. McFarland, Appl. Phys. Lett. **74**, 4046 (1999).

[121] J. W. Schultze and M. M. Lohrengel, Electrochimica Acta **45**, 2499 (2000).

[122] P. A. Redhead, Vacuum **12**, 203 (1962).

[123] K. W. Kolasinski, in *Surface Science: Foundations of Catalysis Nanoscience* (John Wiley & Sons, Chichester, 2009), p. 248.

[124] R. J. Cvetanovic and Y. Amenomiya, Catalysis Reviews **6**, 21 (1972).

[125] A. Peckhaus, Bachelor thesis, Fachbereich Chemie, Universität Duisburg-Essen, 2009.

[126] S. V. Patel, J. L. Gland, and J. W. Schwank, Langmuir **15**, 3307 (1999).

[127] M. Kiskinova, G. Pirug, and H. P. Bonzel, Surface Science **133**, 321 (1983).

[128] H. Steininger, S. Lehwald, and H. Ibach, Surface Science **123**, 264 (1982).

[129] Y. Y. Yeo, L. Vattuone, and D. A. King, The Journal of Chemical Physics **106**, 392 (1997).

[130] A. J. Komrowski, J. Z. Sexton, A. C. Kummel, M. Binetti, O. Weisse, and E. Hasselbrink, Phys. Rev. Lett. **87**, 246103 (2001).

[131] M. Binetti, O. Weisse, E. Hasselbrink, A. J. Komrowski, and A. C. Kummel, Faraday Discussions **117**, 313 (2000).

[132] S. R. Cook, Phys. Rev. (Series I) **18**, 23 (1904).

[133] K. E. Guthe, Phys. Rev. (Series I) **15**, 327 (1902).

[134] B. C. Lai and J. Y. Lee, J. Electrochem. Soc. **146**, 266 (1999).

[135] M. M. Lohrengel, Mater. Sci. Eng. R: **11**, 243 (1993).

[136] K. Shimizu, R. C. Furneaux, G. E. Thompson, G. C. Wood, A. Gotoh, and K. Kobayashi, Oxidation of Metals **35**, 427 (1991).

[137] A. Stoneham, J. Gavartin, and A. Shluger, J. Phys.: Condens. Matter **17**, S2027 (2005).

[138] E. Verwey, Physica **2**, 1059 (1935).

[139] N. Cabrera and N. F. Mott, Rep. Prog. Phys. **12**, 163 (1949).

[140] R. Freer, Journal of Materials Science **15**, 803 (1980).

[141] P. Harrop, Journal of Materials Science **3**, 206 (1968).

[142] M. Huang and K. Hebert, J. Electrochem. Soc. **146**, 3741 (1999).

[143] E. Tan, P. G. Mather, A. C. Perrella, J. C. Read, and R. A. Buhrman, Phys. Rev. B **71**, 161401 (2005).

[144] H. D. Ebinger and J.T. Yates, Jr., Phys. Rev. B **57**, 1976 (1998).

[145] V. Zhukov, I. Popova, and J.T. Yates, Jr., Phys. Rev. B **65**, 195409 (2002).

[146] I. Popova, V. Zhukov, and J.T. Yates, Jr., Phys. Rev. Lett. **89**, 276101 (2002).

[147] A. W. Hassel and D. Diesing, Thin Solid Films **414**, 296 (2002).

[148] R. Prescott and M. J. Graham, Oxidation of Metals **38**, 233 (1992).

[149] J. Schäfer and C. Adkins, J.Phys.: Condens. Matter **3**, 2907 (1991).

[150] M. K. Konkin and J. G. Adler, J. Appl. Phys. **51**, 5450 (1980).

[151] M. K. Konkin and J. G. Adler, J. Appl. Phys. **53**, 5057 (1982).

[152] P. Snijders, L.P.H. Jeurgens, and W.G. Sloof, Surf. Sci. **496**, 97 (2002).

[153] P.C. Snijders, L.P.H. Jeurgens, and W.G. Sloof, Surf. Sci. **589**, 98 (2005).

[154] S. Dobal, R. S. Katiyar, Y. Jiang, R. Guo, and A. S. Bhalla, Journal of Raman spectroscopy **31**, 1061 (2000).

[155] S. Ezhilvalavan and T. Tseng, Journal of Materials Science: Materials in Electronics **10**, 9 (1999).

[156] S. Kamiyama, H. Suzuki, H. Watanabe, A. Sakai, H. Kimura, and J. Mizuki, J. Electrochem. Soc. **141**, 1246 (1994).

[157] S. Byeon and Y. Tzeng, IEEE Transactions on Electronic Devices **37**, 972 (1990).

[158] J. Behler, B. Delley, S. Lorenz, K. Reuter, and M. Scheffler, Phys. Rev. Lett. **94**, 036104 (2005).

[159] J. H. Harding, K. J. W. Atkinson, and R. W. Grimes, J. Am. Ceram. Soc. **86**, 554 (2003).

[160] A. Heuer and K. Lagerlöf, Philosophical Magazine Letters **79**, 619 (1999).

[161] L.P.H. Jeurgens, W.G. Sloof, F.D. Tichelaar, and E.J. Mittemeijer, Thin Solid Films **418**, 89 (2002).

[162] L.P.H. Jeurgens, W.G. Sloof, F.D. Tichelaar, C.G. Borsboom, and E.J. Mittemeijer, Applied Surface Science **144**, 11 (1999).

[163] A. Munoz and J. Bessone, Thin Solid Films **460**, 143 (2004).

[164] C. Gomez-Aleixandre, I. Montero, and J. M. Albella, J. Applied Electrochemistry **16**, 964 (1986).

[165] A. Moehring, M. Pilaski, and M.M. Lohrengel, Ionics **5**, 23 (1999).

[166] D. Diesing, A.W. Hassel, and M.M. Lohrengel, Thin Solid Films **342**, 283 (1999).

[167] L.P.H. Jeurgens, W.G. Sloof, F.D. Tichelaar, and E.J. Mittemeijer, J. Appl. Phys. **92**, 1649 (2002).

[168] A. Wehner, Y. Jeliazova, and R. Franchy, Surface Science **531**, 287 (2003).

[169] D. L. (Ed.), in *CRC Handbook of Chemistry and Physics* (CRC Press Boca Raton London New York Washington DC, London, 1996), Chap. 4, pp. 4–122.

[170] F. S. Ohuchi, R. H. French, and R. V. Kasowski, J. Appl. Phys. **62**, 2286 (1987).

[171] S. Varma, G. S. Chottiner, and M. Arbab, J. Vac. Sci. Technol. **10**, 2857 (1992).

[172] I. Olefjord and A. Nylund, Surface and Interface Analysis **21**, 290 (2004).

[173] M. Handke, C. Paluszkiewicz, and W. Wyrwa, Materials Chemistry **5**, 199 (1980).

[174] D. R. Jennison and T. R. Mattsson, Surf. Sci. **544**, L689 (2003).

[175] K. Gilroy and W. Phillips, Philosophical Magazine B **43**, 735 (1981).

[176] H. Kliem, IEEE Transactions on Insulation **24**, 185 (1989).

[177] J. Chatelet, H. Claassen, D. Gruen, I. Sheft, and R. Wright, Applied Spectroscopy **29**, 185 (1975).

[178] P. J. Chen, M. L. Colaianni, and J. T. Yates, Phys. Rev. B **41**, 8025 (1990).

[179] M. Lee, J. Lee, B. Frederick, and N. Richardson, Surf. Sci. **448**, L 207 (2000).

[180] Y. Oishi and W. D. Kingery, J. Chem. Phys. **33**, 480 (1960).

[181] S. Blonski and S. H. Garofalini, Surf. Sci. **295**, 263 (1993).

[182] S. R. Nagel and S. E. Schnatterly, Phys. Rev. B **9**, 1299 (1974).

[183] E. A. Irene, E. Tierney, and J. Angilello, J. Electrochem. Soc. **129**, 2594 (1982).

[184] Y. M. Zhou, Z. Xie, H. N. Xiao, P. F. Hu, and J. He, Vacuum **83**, 286 (2008).

[185] M. Guziewicz, A. Piotrowska, E. Kaminska, K. Grasza, R. Diduszko, A. Stonert, A. Turos, M. Sochacki, and J. Szmidt, Materials Science and Engineering: B **135**, 289 (2006).

[186] M. Zhang, B. Yang, J. Chu, and T. G. Nieh, Scripta Materialia **54**, 1227 (2006).

[187] M. Zhang, Y. F. Zhang, P. D. Rack, M. K. Miller, and T. G. Nieh, Scripta Materialia **57**, 1032 (2007).

[188] L. Liu, Y. Wang, and H. Gong, Journal of Applied Physics **90**, 416 (2001).

[189] D. Gerstenberg and C. J. Calbick, Journal of Applied Physics **35**, 402 (1964).

[190] P. N. Baker, Thin Solid Films **14**, 3 (1972).

[191] P. N. Baker, Thin Solid Films **8**, R3 (1971).

[192] W. D. Westwood and F. C. Livermore, Thin Solid Films **8**, R1 (1971).

[193] L. G. Feinstein and R. D. Huttemann, Thin Solid Films **20**, 103 (1974).

[194] L. G. Feinstein and R. D. Huttemann, Thin Solid Films **16**, 129 (1973).

[195] L. G. Feinstein and D. Gerstenberg, Thin Solid Films **10**, 79 (1972).

[196] W. Westwood, N. Waterhouse, and P. Wilcox, *Tantalum thin films* (Academic Press, London, Bell Northern Research Ottawa, 1975).

[197] N. O. Nnolim, T. A. Tyson, and L. Axe, Journal of Applied Physics **93**, 4543 (2003).

[198] S. L. Lee, D. Windover, T. M. Lu, and M. Audino, Thin Solid Films **420-421**, 287 (2002).

[199] J. Narayan, V. Bhosle, A. Tiwari, A. Gupta, P. Kumar, and R. Wu, Journal of Vacuum Science Technology A: Vacuum, Surfaces, and Films **24**, 1948 (2006).

[200] J. Zhang, Y. Huai, L. Chen, and J. Zhang, Journal of Vacuum Science & Technology B **21**, 237 (2003).

[201] K. Valleti, A. Subrahmanyam, and S. V. Joshi, Surface and Coatings Technology **202**, 3325 (2008).

[202] N. Schwartz, W. A. Reed, P. Polash, and M. H. Read, Thin Solid Films **14**, 333 (1972).

[203] F. Sajovec, P. M. Meuffels, and T. Schober, Thin Solid Films **219**, 206 (1992).

[204] W. Johnson, JOM Journal of the Minerals, Metals and Materials Society **54**, 40 (2002).

[205] A. Hassel, private communications (2010).

[206] J. D. Torre, G. H. Gilmer, D. L. Windt, R. Kalyanaraman, F. H. Baumann, P. L. O'Sullivan, J. Sapjeta, T. D. de la Rubia, and M. D. Rouhani, Journal of Applied Physics **94**, 263 (2003).

[207] S.-M. Na, I.-S. Park, S.-Y. Park, G.-H. Jeong, and S.-J. Suh, Thin Solid Films **516**, 5465 (2008).

[208] J. H. Mooij, Physica Status Solidi (a) **17**, 521 (1973).

[209] C. C. Tsuei, Phys. Rev. Lett. **57**, 1943 (1986).

[210] A. J. Ahearn, Phys. Rev. **50**, 238 (1936).

[211] R. E. Honig, RCA Review **23**, 567 (1962).

[212] K. Bruder, A. Hassel, K. Stella, D. Diesing, and A. Mardare, in preparation (2011).

[213] K. Stella and D. Diesing, in preparation (2011).

[214] Y. Martin, C. C. Williams, and H. K. Wickramasinghe, Journal of Applied Physics **61**, 4723 (1987).

[215] V. Macagno and J. W. Schultze, Journal of Electroanalytical Chemistry **180**, 157 (1984).

[216] A. Mozalev, A. J. Smith, S. Borodin, A. Plihauka, A. W. Hassel, M. Sakairi, and H. Takahashi, Electrochimica Acta **54**, 935 (2009).

[217] V. Rosenband and A. Gany, Corrosion Science **37**, 1991 (1995).

[218] S. Sankaranarayanan and S. Ramanathan, Physical Review B **78**, 085420 (2008).

[219] G. Reiss, J. Vancea, and H. Hoffmann, Phys. Rev. Lett. **56**, 2100 (1986).

[220] H. Hoffmann and J. Vancea, Thin Solid Films **85**, 147 (1981).

[221] in *CRC handbook of chemistry and physics*, edited by D. Lide (CRC Press Boca Raton London New York Washington DC, London, 1996), Chap. 12, pp. 12–41.

[222] A. Berthault, L. Arles, and J. Matricon, International Journal of Thermophysics **7**, 167 (1986).

[223] G. Pottlacher and A. Seifter, International Journal of Thermophysics **23**, 1281 (2002).

[224] E. H. Sondheimer, Advances in Physics **1**, 1 (1952).

[225] N. Ashcroft and N. Mermin, in *Solid State Physics* (Holt Saunders International Editions, Philadelphia, 1976), Chap. 1, p. 38.

[226] W. B. Jackson et al., Journal of Non-Crystalline Solids **352**, 859 (2006).

[227] H. Prigent, P. Pellen-Mussi, G. Cathelineau, and M. Bonnaure-Mallet, Journal of Biomedical Materials Research **39**, 200 (1998).

[228] M. Knudsen, Annalen der Physik **357**, 105 (1917).

[229] M. Knudsen, Annalen der Physik **353**, 1113 (1916).

[230] T. Matsushima, Progress in Surface Science **82**, 435 (2007).

[231] K. Stella, D. Bürstel, E. Hasselbrink, and D. Diesing, physica status solidi (RRL) –Rapid Research Letters **5**, 68 (2011).

[232] C. J. Wu, P. Soderlind, J. N. Glosli, and J. E. Klepeis, Nature Materials **8**, 223 (2009).

[233] D. Errandonea, Nature Materials **8**, 170 (2009).

[234] K. Hieber and N. M. Mayer, Thin Solid Films **90**, 43 (1982).

[235] M. A. Park, K. Savran, and Y. J. Kim, physica status solidi (b) **237**, 500 (2003).

[236] S. S. Li and W. R. Thurder, Solid-State Electronics **20**, 609 (1977).

[237] K. Andersson, K. Reichert, and R. Wolf, *Tantalum and Tantalum Compounds* (Ullmann's Encyclopedia of Industrial Chemistry Wiley VCH, Weinheim, 2000).

[238] S. P. Murarka, D. B. Fraser, W. S. Lindenberger, and A. K. Sinha, Journal of Applied Physics **51**, 3241 (1980).

[239] D. A. Robins, Philosophical Magazine **3**, 313 (1958).

[240] D. L. Kwong, Thin Solid Films **121**, 43 (1984).

[241] X. Q. Cheng, R. S. Wang, X. J. Tang, and B. X. Liu, Journal of Alloys and Compounds **363**, 236 (2004).

[242] K.-C. Tsai, W.-F. Wu, C.-G. Chao, and C.-P. Kuan, Journal of The Electrochemical Society **153**, G492 (2006).

[243] R. Dreiner, Journal of The Electrochemical Society **111**, 27 (1964).

[244] C. L. Platt, B. Dieny, and A. E. Berkowitz, Journal of Applied Physics **81**, 5523 (1997).

[245] D. R. Jennison, P. A. Schultz, and J. P. Sullivan, Phys. Rev. B **69**, 041405 (2004).

[246] V. V. Afanas'ev and A. Stesmans, Journal of Applied Physics **102**, 081301 (2007).

[247] V. Shvets *et al.*, Journal of Non-Crystalline Solids **354**, 3025 (2008).

[248] K. Kukli, J. Aarik, A. Aidla, O. Kohan, T. Uustare, and V. Sammelselg, Thin Solid Films **260**, 135 (1995).

[249] J. Robertson and C. W. Chen, Applied Physics Letters **74**, 1168 (1999).

[250] L. Miao, P. Jin, K. Kaneko, A. Terai, N. Nabatova-Gabain, and S. Tanemura, Applied Surface Science **212-213**, 255 (2003).

[251] S. Maeng, L. Axe, T. Tyson, and A. Jiang, Journal of The Electrochemical Society **152**, B60 (2005).

[252] J. W. Schultze and V. A. Macagno, Electrochimica Acta **31**, 355 (1986).

[253] W. Schmickler and J. Schultze, in *Electron Transfer Reactions on Oxide Covered Metal Electrodes*, edited by J. O. Bockris, B. E. Conway, and R. E. White (Plenum Press New York, New York, 1986), Vol. 17, p. 357.

[254] D. Hunkel, M. Marso, R. Butz, R. Arens-Fischer, and H. Lüth, Materials Science and Engineering B **69-70**, 100 (2000).

[255] D. Diesing, G. Kritzler, M. Stermann, D. Nolting, and A. Otto, J. Solid State Electrochemistry **7**, 389 (2003).

[256] J. Blakemore, in *Solid State Physics* (Cambridge University Press, Cambridge New York Melbourne, 1985), p. 152.

[257] S. H. Overbury, P. A. Bertrand, and G. A. Somorjai, Chemical Reviews **75**, 547 (1975).

[258] C. Bombis, A. Emundts, M. Nowicki, and H. P. Bonzel, Surface Science **511**, 83 (2002).

[259] U. Breuer and H. P. Bonzel, Surface Science **273**, 219 (1992).

[260] J. C. Heyraud and J. J. Metois, Acta Metallurgica **28**, 1789 (1980).

[261] M. McLean and H. Mykura, Surface Science **5**, 466 (1966).

[262] L. Z. Mezey and J. Giber, Surface Science **117**, 220 (1982).

[263] E. Bauer, Zeitschrift für Kristallographie **110**, 423 (1958).

[264] J. H. van der Merwe and E. Bauer, Phys. Rev. B **39**, 3632 (1989).

[265] F. Cosandey and T. E. Madey, Surface Review and Letters **8**, 73 (2001).

[266] in *CRC Materials Science and Engineering Handbook* (CRC Press Boca Raton London New York Washington DC, London, 2001), p. 253.

[267] in *CRC Materials Science and Engineering Handbook* (CRC Press Boca Raton London New York Washington DC, London, 2001), p. 254.

[268] R. Tadmor, Langmuir **20**, 7659 (2004).

[269] A. Grossmann, W. Erley, J. B. Hannon, and H. Ibach, Phys. Rev. Lett. **77**, 127 (1996).

[270] A. Grossmann, W. Erley, J. B. Hannon, and H. Ibach, Phys. Rev. Lett. **78**, 3587 (1997).

[271] W. Haiss, Reports on Progress in Physics **64**, 591 (2001).

[272] D. Edwards, International Journal of Heat and Mass Transfer **25**, 815 (1982).

[273] S. Pepper, J.Opt.Soc.Am. **60**, 805 (1970).

[274] J. L. Ord and W. P. Wang, Journal of The Electrochemical Society **130**, 1809 (1983).

[275] J. L. Ord, D. J. D. Smet, and D. J. Beckstead, Journal of The Electrochemical Society **136**, 2178 (1989).

[276] A. G. Revesz, J. H. Reynolds, and J. F. Allison, Journal of The Electrochemical Society **123**, 889 (1976).

[277] E. Franke, C. L. Trimble, M. J. DeVries, J. A. Woollam, M. Schubert, and F. Frost, Journal of Applied Physics **88**, 5166 (2000).

[278] B. R. Cooper, H. Ehrenreich, and H. R. Philipp, Phys. Rev. **138**, A494 (1965).

[279] H. V. Nguyen, I. An, and R. W. Collins, Phys. Rev. B **47**, 3947 (1993).

[280] L.-J. Meng, V. Teixeira, H. N. Cui, F. Placido, Z. Xu, and M. P. dos Santos, Applied Surface Science **252**, 7970 (2006).

[281] Y. Wouters, A. Galerie, and J.-P. Petit, Journal of The Electrochemical Society **154**, C587 (2007).

[282] P. H. P. Koller, H. J. M. Swagten, W. J. M. de Jonge, H. Boeve, and R. Coehoorn, Applied Physics Letters **84**, 4929 (2004).

[283] P. H. P. Koller, F. W. M. Vanhelmont, H. Boeve, R. Coehoorn, and W. J. M. de Jonge, J. Appl. Phys. **93**, 8549 (2003).

[284] R. Cataliotti, Journal of Physics C: Solid State Physics **7**, 3467 (1974).

[285] R. Franchy, Surf. Sci. Rep. **38**, 195 (2000).

[286] F. L. Schuermeyer, C. R. Young, and J. M. Blasingame, Journal of Applied Physics **39**, 1791 (1968).

[287] F. L. Schuermeyer and J. A. Crawford, Applied Physics Letters **9**, 317 (1966).

[288] R. M. Handy, Journal of Applied Physics **37**, 4620 (1966).

[289] R. H. Fowler, Phys. Rev. **38**, 45 (1931).

[290] V. V. Afanas'ev, M. Houssa, A. Stesmans, and M. M. Heyns, Journal of Applied Physics **91**, 3079 (2002).

[291] Z. Rotenberg, Y. Prischepa, and Y. Pleskov, J. Electroanal.Chem. **56**, 345 (1974).

[292] Y. V. Pleskov and Z. A. Rotenberg, Journal of Electroanalytical Chemistry and Interfacial Electrochemistry **94**, 1 (1978).

[293] Y. V. Gurevich, Y. V. Pleskov, and Z. A. Rothenberg, *Photoelectrochemistry* (Consultants Bureau, Singapore, 1980).

[294] Z. A. Rotenberg, N. V. Gromova, and V. E. Kazarinov, Journal of Electroanalytical Chemistry and Interfacial Electrochemistry **204**, 281 (1986).

[295] R. H. Fowler and L. Nordheim, Proc. Roy. Soc. **119**, 173 (1928).

[296] L. Nordheim, Zeitschrift für Physik **46**, 833 (1928).

[297] G. Wentzel, Zeitschrift für Physik A Hadrons and Nuclei **38**, 518 (1926).

[298] H. A. Kramers, Zeitschrift für Physik A Hadrons and Nuclei **39**, 828 (1926).

[299] L. Brillouin, Comptes Rendus **183**, 24 (1926).

[300] P. Thissen, B. Schindler, D. Diesing, and E. Hasselbrink, New Journal of Physics **12**, 113014 (2010).

[301] S. John, C. Soukoulis, M. H. Cohen, and E. N. Economou, Phys. Rev. Lett. **57**, 1777 (1986).

[302] F. Urbach, Phys. Rev. **92**, 1324 (1953).

[303] J. P. Masse, H. Szymanowski, O. Zabeida, A. Amassian, J. E. Klemberg-Sapieha, and L. Martinu, Thin Solid Films **515**, 1674 (2006).

[304] R. M. Fleming et al., Journal of Applied Physics **88**, 850 (2000).

[305] J. R. Nesbitt and A. F. Hebard, Physical Review B (Condensed Matter and Materials Physics) **75**, 195441 (2007).

[306] A. K. Jonscher, Journal of Physics D: Applied Physics **32**, R57 (1999).

[307] J. Curie, Annales de chemie et de physique **17**, 385 (1889).

[308] E. R. v. Schweidler, Annalen der Physik **329**, 711 (1907).

[309] P. M. Kumar, S. Badrinarayanan, and M. Sastry, Thin Solid Films **358**, 122 (2000).

[310] J. M. Sanz, L. Soriano, P. Prieto, G. Tyuliev, C. Morant, and E. Elizalde, Thin Solid Films **332**, 209 (1998).

[311] K. Gundlach, J. Appl. Phys **44**, 5005 (1973).

[312] C. B. Duke, *Tunneling in Solids* (Academic Press, 111 Fifth Avenue New York, 1969), p. 62.

[313] E. Nicollian and J. Brews, *MOS (Metal Oxide Semiconductor) Physics and Technology* (John Wiley and Sons, New York Chichester Brisbane, 1982).

[314] H. C. Card and E. H. Rhoderick, Journal of Physics D: Applied Physics **4**, 1589 (1971).

[315] H. C. Card and E. H. Rhoderick, Journal of Physics D: Applied Physics **4**, 1602 (1971).

[316] A. Waxman, J. Shewchun, and G. Warfield, Solid-State Electronics **10**, 1187 (1967).

[317] J. Shewchun, A. Waxman, and G. Warfield, Solid-State Electronics **10**, 1165 (1967).

[318] A. H. M. Shousha, Phys. Lett. A **92**, 293 (1982).

[319] D. R. Lillington and W. G. Townsend, Applied Physics Letters **28**, 97 (1976).

[320] L. M. Terman, Solid-State Electronics **5**, 285 (1962).

[321] H. Aguas, A. Goullet, L. Pereira, E. Fortunato, and R. Martins, Thin Solid Films **451-452**, 361 (2004).

[322] C.-H. Chen, C.-C. Hong, and J.-G. Hwu, Journal of The Electrochemical Society **149**, G362 (2002).

[323] Y.-P. Lin and J.-G. Hwu, Journal of The Electrochemical Society **151**, G853 (2004).

[324] C.-Y. Wang and J.-G. Hwu, Journal of The Electrochemical Society **156**, H181 (2009).

[325] D. M. Fleetwood, P. S. Winokur, J. R. A. Reber, T. L. Meisenheimer, J. R. Schwank, M. R. Shaneyfelt, and L. C. Riewe, Journal of Applied Physics **73**, 5058 (1993).

[326] R. Williams, Phys. Rev. **140**, A569 (1965).

[327] A. M. Goodman, Phys. Rev. **144**, 588 (1966).

[328] W. Daum, Applied Physics A: Materials Science & Processing **87**, 451 (2007).

[329] F. Yakuphanoglu, Sensors and Actuators A: Physical **147**, 104 (2008).

[330] B. R. Cuenya, H. Nienhaus, and E. W. McFarland, Physical Review B (Condensed Matter and Materials Physics) **70**, 115322 (2004).

[331] R. E. Oakley and M. Pepper, Phys. Lett. A **41**, 87 (1972).

[332] A. Muñoz and M.M. Lohrengel, Journal of Solid State Electrochemistry **6**, 513 (2002).

[333] A. Muñoz, A. Moehring, and M.M. Lohrengel, Electrochimica Acta **47**, 2751 (2002).

[334] E. Nicollian and A. Goetzberger, The Bell System Technical Journal **46**, 1055 (1967).

[335] T. Greber, Surface Science Reports **28**, 1 (1997).

[336] L. B. Thomsen, G. Nielsen, S. Vendelbo, M. Johansson, O. Hansen, and I. Chorkendorff, Physical Review B (Condensed Matter and Materials Physics) **76**, 155315 (2007).

[337] L. B. Thomsen, G. Nielsen, S. B. Vendelbo, M. Johansson, O. Hansen, and I. Chorkendorff, Journal of Vacuum Science and Technology B: Microelectronics and Nanometer Structures **27**, 562 (2009).

[338] G. Nielsen, L. B. Thomsen, M. Johansson, O. Hansen, and I. Chorkendorff, Applied Surface Science **255**, 7657 (2009).

[339] J. F. Ziegler, J. P. Biersack, and M. D. Ziegler, *SRIM - The Stopping and Range of Ions in Matter* (SRIM Co., Lulu Press Co Morrisville, 2008).

[340] U. Hagemann, private communications (2010).

[341] N. Terada, T. Haga, N. Miyata, K. Moriki, M. Fujisawa, M. Morita, T. Ohmi, and T. Hattori, Phys. Rev. B **46**, 2312 (1992).

[342] M. Hiroshima, T. Yasaka, S. Miyazaki, and M. Hirose, Japanese Journal of Applied Physics **33**, 395 (1994).

[343] M. Shur, in *Physcis of Semiconductor Devices* (Prentice Hall, Englewood Cliffs, New Jersey 07632, 1990), pp. 343–352.

[344] W. A. Yager, Physics **7**, 434 (1936).

[345] R. Lindner, The Bell System Technical Journal **41**, 803 (1962).

[346] E. Nicollian and J. Brews, in *MOS (Metal Oxide Semiconductor) Physics and Technology* (John Wiley and Sons, New York Chichester Brisbane, 1982), p. 424.

[347] E. Nicollian and J. Brews, in *MOS (Metal Oxide Semiconductor) Physics and Technology* (John Wiley and Sons, New York Chichester Brisbane, 1982), p. 64.

[348] E. Nicollian and J. Brews, in *MOS (Metal Oxide Semiconductor) Physics and Technology* (John Wiley and Sons, New York Chichester Brisbane, 1982), p. 427.

[349] E. Nicollian and J. Brews, in *MOS (Metal Oxide Semiconductor) Physics and Technology* (John Wiley and Sons, New York Chichester Brisbane, 1982), p. 426.

[350] P. A. Anderson, Phys. Rev. **115**, 553 (1959).

[351] F. G. Allen and G. W. Gobeli, Phys. Rev. **127**, 150 (1962).

[352] E. O. Kane, Phys. Rev. **127**, 131 (1962).

[353] L. B. Thomsen, Ph.D. thesis, Center for Individual Nanoparticle Functionality, Department of Physics, Technical University of Denmark, 2009.

[354] F. Giustino, A. Bongiorno, and A. Pasquarello, Applied Physics Letters **86**, 192901 (2005).

[355] S. Wakui, J. Nakamura, and A. Natori, Japanese Journal of Applied Physics **46**, 3261 (2007).

[356] S. Wakui, J. Nakamura, and A. Natori, Journal of Vacuum Science & Technology B **24**, 1992 (2006).

[357] S. Kar and W. E. Dahlke, Applied Physics Letters **18**, 401 (1971).

[358] H. Deuling, E. Klausmann, and A. Goetzberger, Solid-State Electronics **15**, 559 (1972).

[359] J. Kadlec, Physics Reports **26**, 69 (1976).

[360] R. Tohmon, H. Mizuno, Y. Ohki, K. Sasagane, K. Nagasawa, and Y. Hama, Phys. Rev. B **39**, 1337 (1989).

[361] S. Shevenock, S. Fonash, and J. Geneczko, Electron Devices Meeting, 1975 International **21**, 211 (1975).

[362] R. Singh and J. Shewchun, Applied Physics Letters **28**, 512 (1976).

[363] E. P. O'Reilly and J. Robertson, Phys. Rev. B **27**, 3780 (1983).

[364] D. Papaconstantopoulos, in *Handbook of the Band Structure of elemental solids* (Plenum Press, New York, 1986), p. 202.

[365] D. Papaconstantopoulos, in *Handbook of the Band Structure of elemental solids* (Plenum Press, New York, 1986), p. 198.

[366] A. Goldmann, R. Matzdorf, and F. Theilmann, Surf. Sci. **414**, L 932 (1998).

[367] W. S. Fann, R. Storz, H. W. K. Tom, and J. Bokor, Phys. Rev. B **46**, 13592 (1992).

[368] M. Lindenblatt, E. Pehlke, A. Duvenbeck, B. Rethfeld, and A. Wucher, Nucl. Instrum. Methods Phys. Res. B **246**, 333 (2006).

[369] J. Vanhellemont, E. Simoen, A. Kaniava, M. Libezny, and C. Claeys, Journal of Applied Physics **77**, 5669 (1995).

[370] S. Meyer, C. Heuser, D. Diesing, and A. Wucher, Physical Review B (Condensed Matter and Materials Physics) **78**, 035428 (2008).

[371] G. Lakits, F. Aumayr, M. Heim, and H. Winter, Phys. Rev. A **42**, 5780 (1990).

[372] M. Prietsch, Physics Reports-Review Section of Physics Letters **253**, 163 (1995).

[373] M. J. Gordon, X. Qin, A. Kutana, and K. P. Giapis, Journal of the American Chemical Society **131**, 1927 (2009).

[374] J. Mace, M. J. Gordon, and K. P. Giapis, Physical Review Letters **97**, 257603 (2006).

[375] C. L. Quinteros, T. Tzvetkov, and D. C. Jacobs, The Journal of Chemical Physics **113**, 5119 (2000).

[376] W. R. Williams, C. M. Marks, and L. D. Schmidt, Journal of Physical Chemistry **96**, 5922 (1992).

[377] V. J. Kwasniewski and L. D. Schmidt, Journal of Physical Chemistry **96**, 5931 (1992).

[378] S. T. Ceyer, W. L. Guthrie, T. H. Lin, and G. A. Somorjai, The Journal of Chemical Physics **78**, 6982 (1983).

[379] D. Diesing, D. Kovacs, K. Stella, and C. Heuser, Nuclear Instruments and Methods in Physics Research Section B: Beam Interactions with Materials and Atoms **In Press, Corrected Proof**, .

[380] K. Tschersich, U. Littmark, and W. Beyer, Thin Solid Films **515**, 464 (2006).

[381] W. Umrath, in *Fundamentals of Vacuum Technology* (Leybold, Köln, 1998), p. 83.

[382] M. Wilde and K. Fukutani, Physical Review B (Condensed Matter and Materials Physics) **78**, 115411 (2008).

[383] E. Nowicka, Z. Wolfram, and R. Dus, Surface Science **247**, 248 (1991).

[384] K. Nobuhara, H. Kasai, W. Dino, H. Nakanishi, and A. Okiji, Japanese Journal of Applied Physics **42**, 4630 (2003).

[385] K. Bonhoeffer, Ergebnisse der exakten Naturwissenschaften **6**, 201 (1927).

[386] W. E. Lamb and R. C. Retherford, Physical Review **79**, 549 (1950).

[387] W. Göpel and H. D. Wiemhöfer, in *Statistische Thermodynamik* (Spektrum Akademischer Verlag GmbH, Heidelberg Berlin, 2000), p. 578.

[388] W. Göpel and H. D. Wiemhöfer, in *Statistische Thermodynamik* (Spektrum Akademischer Verlag GmbH, Heidelberg Berlin, 2000), p. 578.

[389] W. Göpel and H. D. Wiemhöfer, in *Statistische Thermodynamik* (Spektrum Akademischer Verlag GmbH, Heidelberg Berlin, 2000), p. 577.

[390] W. Göpel and H. D. Wiemhöfer, in *Statistische Thermodynamik* (Spektrum Akademischer Verlag GmbH, Heidelberg Berlin, 2000), p. 574.

[391] W. Göpel and H. D. Wiemhöfer, in *Statistische Thermodynamik* (Spektrum Akademischer Verlag GmbH, Heidelberg Berlin, 2000), p. 174.

[392] W. Göpel and H. D. Wiemhöfer, in *Statistische Thermodynamik* (Spektrum Akademischer Verlag GmbH, Heidelberg Berlin, 2000), p. 175.

[393] W. Demtröder, private communications (2011).

[394] G. O. Sitz, private communications (2011).

[395] F. D. Shields and K. P. Lee, The Journal of Chemical Physics **40**, 737 (1964).

[396] F. D. Shields and K. P. Lee, The Journal of the Acoustical Society of America **35**, 251 (1963).

[397] F. D. Shields, The Journal of the Acoustical Society of America **32**, 180 (1960).

[398] F. R. S. G. Herzberg, in *Molecular spectra and molecular structure I. Spectra of diatomic molecules* (Robert E. Krieger Publishing Company, Malabar, Florida, 1989), p. 560.

[399] F. R. S. G. Herzberg, in *Molecular spectra and molecular structure I. Spectra of diatomic molecules* (Robert E. Krieger Publishing Company, Malabar, Florida, 1989), p. 532.

[400] F. R. S. G. Herzberg, in *Molecular spectra and molecular structure I. Spectra of diatomic molecules* (Robert E. Krieger Publishing Company, Malabar, Florida, 1989), p. 92.

[401] E. Santos, A. Lundin, K. Potting, P. Quaino, and W. Schmickler, Physical Review B (Condensed Matter and Materials Physics) **79**, 235436 (2009).

[402] J. Schulze and A.W. Hassel, in *Passivity of metals, alloys and semiconductors*, Vol. 4 of *Encyclopedia of Electrochemistry*, edited by M. Stratmann (Wiley VCH, Weinheim, 2003), Chap. 3.2, p. 216.

[403] P. Wissmann, *Thin metal films and gas chemisorption, Studies in surface science and catalysis; 32* (Elsevier, Amsterdam, 1987).

[404] H. Kanter, Phys. Rev. B **1**, 522 (1970).

[405] J. C. Ashley, C. J. Tung, and R. H. Ritchie, Surface Science **81**, 409 (1979).

6 Further experimental results planned for publication

6.1 Temperature dependence of internal photoemission in MIM sensors

In this internal photoemission study I use tantalum–tantalum oxide–gold (MIM) sensors. They consist of a 20 nm thick, 2 mm wide and 20 mm long amorphous tantalum back electrode, which was deposited with an electron beam evaporator under Ultra High Vacuum conditions on ethanol cleaned glass substrates, an electrochemically formed tantalum oxide of 4 nm and a gold top layer of 20 nm thickness, which was thermally evaporated (see sections 4.4 and 4.3). The samples are mounted on a heatable (tungsten filament) and coolable (He-Kryostat) sample holder in an Ultra High Vacuum chamber. The bottom and the top electrode of the MIM sensor are connected to a potentiostat, which is used to apply a desired bias voltage to the sensor in the bias voltage dependent internal photoemission studies and measure the internal photoemission current.

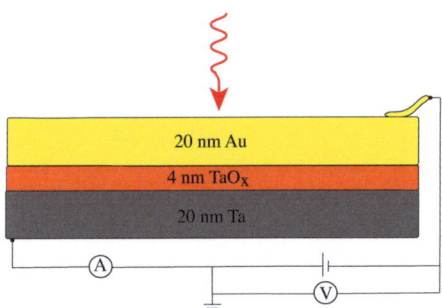

Figure 6.1: Schematic set-up of the tantalum–tantalum oxide–gold (MIM) detector during irradiation with light.

6.1 Temperature dependence of internal photoemission in MIM sensors

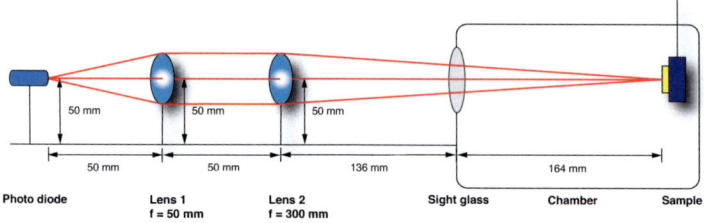

Figure 6.2: Schematic drawing of the used lens system to focus the laser beam on the sample in the UHV chamber

Irradiation of the samples (see figure 6.1) was carried out with manually chopped laser diodes (635 nm, 808 nm, 904 nm and 980 nm respectively 1.95 eV, 1.53 eV, 1.37 eV and 1.27 eV) and a lens system to focus the laser beam on the sample in the UHV chamber under normal incidence (see figure 6.2). The temperature dependent photon-induced currents, further called internal photoemission, were always measured on the cooling down cycle to prevent any heating effects or irradiation by the tungsten filament from disturbing the measurements. The measured current is defined as the net number of flowing electrons into the bottom tantalum electrode per unit time. A negative current then means a net current of electrons flowing from the bottom into the top metal electrode. The photon-induced electron emission is characterized by a yield γ, defined as the average number of detected electrons induced by one incoming photon. Thereby the calibration of the used laser diodes was performed by guiding the laser beam through the same lenses and the same chamber window onto a silicon photo diode with known sensitivity. Additionally, I use the normalized photoyield, which is the photoyield at the temperature T normalized by the photoyield at 61 K (lowest reachable temperature).

In figure 6.3 the measured temperature dependence of the normalized internal photoemission yield of a tantalum–tantalum oxide–gold sensor with an applied voltage of 0 V is shown for the four used photon energies. A pronounced temperature dependence of the internal photoemission can be observed. A signal increase of around one order of magnitude is measured for the internal photoemission for the 635 nm (1.95 eV) excitation wavelength (excitation energy higher than the tantalum–tantalum oxide barrier of 1.7 eV). When the internal photoemission is carried out with an illumination with 808 nm (1.53 eV) excitation wavelength (slightly smaller excitation energy than the tantalum–tantalum oxide barrier) the signal increases by a factor of 570 in the temperature range between 61 and 415 K.

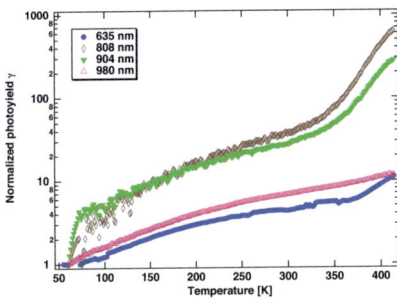

Figure 6.3: Measured temperature dependence of the internal photoemission yield of a tantalum–tantalum oxide–gold sensor with an applied voltage of 0 V. The photoyield is normalized at 61 K.

The excitations with energies much smaller than the tantalum–tantalum oxide barrier, as 904 nm wavelength (1.37 eV) and 980 nm wavelength (1.27 eV) show an increase of the signals by a factor of 270, respectively 10 in the temperature range between 61 and 415 K

To check whether such a pronounced temperature dependence of the internal photoemission is understandable, a simple model based calculation is carried out. In this simple model, excitations in a tantalum–tantalum oxide–gold system are simulated with a laser energy of 1 eV. Barrier heights are 1 eV at the tantalum–tantalum oxide, 1.7 eV at the tantalum oxide–gold interface. 4 eV is assumed for the band gap of the tantalum oxide. To explain the fundamentals of this calculation an energy level diagram of a tantalum–tantalum oxide–gold sensor under illumination with $h \cdot \nu = 1\,\text{eV}$ is shown as figure 6.4. Such an illumination with $h \cdot \nu = 1\,\text{eV}$ shifts a part of the ground state electron distribution of the gold $\Delta E = 1\,\text{eV}$ upwards (marked with a red rectangle). This part is the source term for the photon-induced current. The simplified photoelectron distribution around $\pm 0.5\,\text{eV} + E_\text{F} + h \cdot \nu$ is added for 300 K and 500 K as illustrated in the inset on the right side. One can clearly see that there is only a weak change with temperature. At a first glance one would not expect a big change of the internal photoemission due to this small change of the photo electron distribution. But one should have a look at the photoelectron distribution that is transported through the oxide according to $f(E) \cdot T(E)$, where $T(E)$ is the tunnel probability calculated in **Wentzel-Kramers-Brillouin** (WKB) approximation. This photoelectron distribution after the transport is shown for 300 K and 500 K around $\pm 1\,\text{eV} + E_\text{F} + h \cdot \nu$ in the inset on the left side of figure 6.4. Here one can denote a dramatic temperature dependence of the photoelectron distribution. The current for 500 K is more than two orders of magnitude larger than for 300 K. Due to the folding of the photo electron distribution with the tunnel probability a maximum in the distribution appears. This current maximum is shifted from $\approx 0.3\,\text{eV} + E_F + h \cdot \nu$ (300 K) to $\approx 0.7\,\text{eV} + E_F + h \cdot \nu$ (500 K). So the transported charge carriers show different excess energies as well.

Thus one does not have to resort to other effects like a changed barrier due to e.g. ionic conduction through the oxide or even a complete destruction of the barrier. The latter one can be excluded due to the procedure of the experiment where the higher temperatures were measured first. Additionally a destruction of the barrier or ionic conduction through the oxide should be observable in a study of the temperature dependence of current voltage curves as well, but that is not the case. The results of the mentioned calculations regarding the temperature dependence of the internal photoemission are plotted in figure 6.5. A signal increase of around four orders of magnitude is calculated over this range. Thus even with such a simple model one can clearly demonstrate that a big temperature dependence of the internal photoemission currents is an intrinsic property of carrier transport through the edge of a tunnel barrier. Temperature induced barrier changes are obviously not necesary to discuss such effects.

Extending these arguments regarding the temperature dependence of the internal photoemission to chemicurrent studies, one would expect a very strong temperature dependence of the chemicurrent as well.

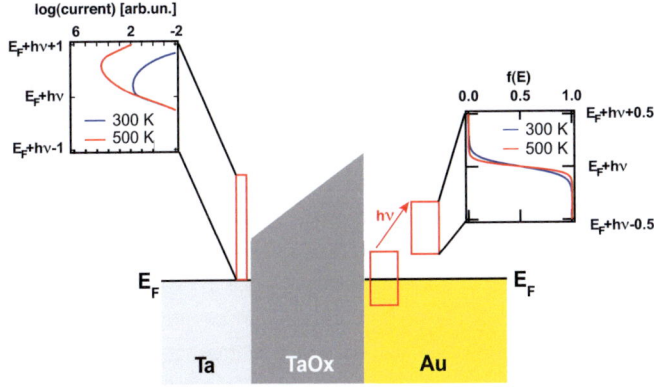

Figure 6.4: Energy level diagram of a tantalum–tantalum oxide–gold sensor under illumination with $h \cdot \nu = 1\,\text{eV}$. Simplified photoelectron distribution which may be represented by a shift $\Delta E = 1\,\text{eV}$ of the ground state electron distribution (inset on the right side). Inset on the left side shows the photoelectron distribution transported through the oxide ($f(E) \cdot T(E)$, where $T(E)$ is the tunnel probability in WKB approximation).

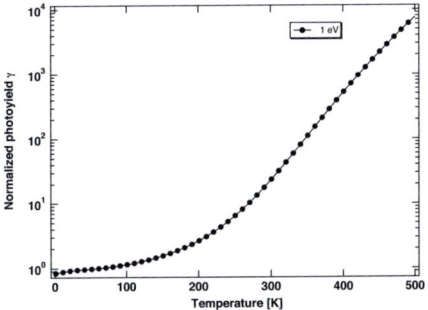

Figure 6.5: Calculated temperature dependence of the internal photoemission yield for a tantalum–tantalum oxide–gold sensor, normalized to 61 K. Illuminated with a photon energy of 1 eV.

6.2 Temperature dependence of internal photoemission in MIS sensors

In this section I will present the temperature dependence of internal photoemission in **stepped-m**etal–**i**nsulator–**s**emiconductor (**stepped-MIS**) chemoelectronic sensors.

In the chapter before we showed the temperature dependence of the internal photoemission in MIM sytems. The photoyield always increased with increasing temperature. This could be explained by the strongly increasing tunnel probability for already thermally excited electrons, since these electrons gain further excess energy by the photoabsorption and thus tunnel through the upper edge of the tunnel barrier. It could be shown that an increasing photocurrent with increasing sensor temperature is typical for systems, where the transport of excited carriers is dominated by tunneling through a barrier.

Studying the internal photoemission in MIS systems, we concluded that the main photoabsorption process occurs in the semiconductor. The high photoyield in MIS systems was explained by the transport and separation of excited holes and electrons in the space charge layers and a nearly barrier less transport through the oxide (compare section 4.8 and figure 4.56). If this model is correct one would assume that a lower temperature is helpful for the internal photoemission in MIS systems. With a lower temperature scattering processes in the space charge layer would not disturb the hot carrier transport. And if the influence of the barrier is low, one should not have this temperature assisted tunneling of excited carriers, as we observed in MIM systems.

So, the study of the temperature dependence of the internal photoemission can be seen as a litmus test for our photo emission model for our MIS systems.

Therefore the temperature dependence of the internal photoemission of a gold–MIS chemoelectronic sensors was studied. This MIS system consists of a 20 × 10 mm^2 large piece of an n-type Si(111) wafer (7.5 Ω·cm), an electrochemically prepared silicon oxide of different thicknesses (1 nm and 4 nm), and a 20 nm thick thermally evaporated gold top layer (compare sections 4.7 and 4.8)

The samples are mounted on a heatable (tungsten filament) and coolable (He-Kryostat) sample holder in an Ultra High Vacuum chamber. The bottom (silicon) and the top electrode (gold) of the stepped-MIS sensor are connected to a potentiostat, which is used to measure the internal photoemission currents at various temperatures. The top metal is contacted at the thicker oxide area to avoid any extrusion of the contacts.

Irradiation of the samples was carried out analogously to section 6.1 (compare figure 6.1) with manually chopped laser diodes (635 nm, 808 nm, 904 nm and 980 nm respectively 1.95 eV, 1.53 eV, 1.37 eV and 1.27 eV) and a lens geometry to focus the laser beam on the thinner oxide area of the MIS samples in the UHV chamber (see figure 6.2). These internal photoemission studies were always measured on the cooling down cycle to exclude any heating effects or irradiation by the tungsten filament disturbing the measurements. In the following, the internal photoemission current is counted positive if a net electron current is flowing from gold or platinum to the silicon. The same sign convention holds for the internal photoemission yield. The photon-induced electron emission current is characterized by a yield γ, defined as the average number of detected electrons induced by one incoming photon (with the same calibrated laser diodes as in section 6.1). Additionally I use the normalized photoyield, which is the photoyield at the temperature T normalized by the photoyield at 68 K.

In figure 6.6 the measured temperature dependence of the normalized internal photoemission yield of a silicon–silicon oxide(1 nm / 4nm)–gold MIS sample (irradiating the 1 nm part) is presented. One can observe a pronounced temperature dependence of the internal photoemission yield. For the gold-MIS sensor the photoemission yield is around 1 order of magnitude higher for lower temperatures and shows quite similar slopes for all probed excitation energies. (For platinum-MIS sensors quite similar results are obtained, but not shown here.) So MIS sensors show a temperature dependence of the internal photoemission the other way round than MIM sensors.

Thus one is able to verify the essence of section 4.8 by this temperature dependence study of the internal photoemission yield. Therefore when temperature dependent chemicurrent studies are performed, one would expect a pronounced temperature dependence as well.

6.2 Temperature dependence of internal photoemission in MIS sensors

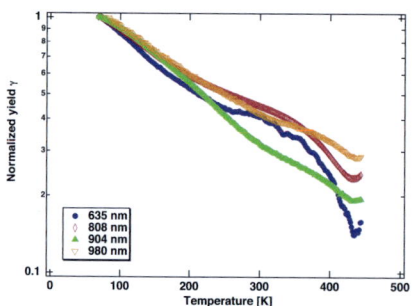

Figure 6.6: Temperature dependence of the internal photoemission for four different wavelength for a silicon–silicon oxide(1 nm / 4 nm)–gold stepped-MIS (irradiating the 1 nm part). Normalized at 68 K.

6.3 About the stability of thin Al–AlOx and Ta–TaOx interfaces

6.3.1 Introduction

Thin tantalum and aluminum films are an intensively studied field in the last years as they can be used for electronic sensors [7,14,15,85,107,184–187]. They can be easily oxidized in an electrochemical droplet cell [82,85,86] (compare section 4.4). However, the question is how stable the remaining metal film is when it is oxidized in such a way that only a few layers remain unoxidized.

In earlier studies of aluminum films I saw that the underlying aluminum metal film is not stable [84] (compare section 4.3), whereas tantalum seems to be stable. Here a comparison between the non stable aluminum–aluminum oxide interface and a tantalum–tantalum oxide interface is carried out. One open question is, up to what minimal thickness of the underlying metal is an electrochemical oxidation possible, and how stable is its metal–metal oxide interface. What values does the specific electrical resistivity show, when only a few layers metal are present? To answer these questions we performed metal–metal oxide interface studies by measuring the electrical resistivity, analogously to section section 4.3 [84], which enable a measurement of the resistance of the oxide's underlying metal.

6.3.2 Experimental

In a previous section 4.4, [82] I showed how thin amorphous tantalum films can be prepared by small e–beam evaporators. These homogeneous films have only a thickness variation of some nm. The aluminum films (as well 10 nm thick 2 mm wide and 2 cm long) were evaporated from a thermal source containing a tungsten basket in a high vacuum chamber on microscope glass slides. During deposition the substrates were kept at room temperature. Slow deposition rates were chosen in both cases to ensure a good homogeneity of the metal films (compare sections 4.3, 4.4 [82,84]). These metal films are thinned by an electrochemical oxidation procedure where a part of the metal is used to form the metal oxide. This electrochemical oxidation was carried out in an electrolytic droplet cell [85,86] (compare section 4.4) using an acetate buffer electrolyte which minimizes parallel corrosion processes during oxidation [87,402]. A classical, waterproof non conductive, varnish was used to define the oxidized area to 0.04 cm^2. After rinsing with MilliQ water and drying the samples in a nitrogen stream, the

6.3 About the stability of thin Al–AlOx and Ta–TaOx interfaces

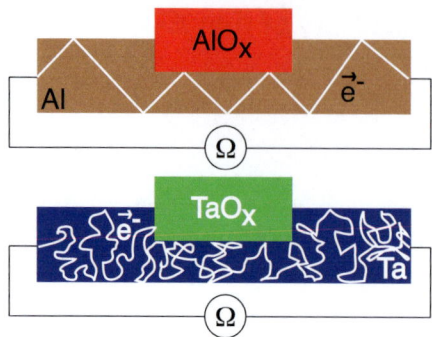

Figure 6.7: Schematic drawing of the samples and the electrical measurement set-up. The metal and its electrochemically subsequent heightened oxide (via consuming a bit of the underlying metal) are sketched as well as possible electron pathways in the metals.

samples are transferred to an electrical resistance measurements unit described earlier [84] (see section 4.3) and sketched in figure 6.7, whereby we probe the oxide's underlying metal resistance. This procedure is repeated and diverse oxidation potentials are applied, and each time the samples are transferred to the electrical measurements unit.

6.3.3 Discussion

6.3.3.1 AFM studies of the Al and Ta thin films on glass

The evaporated metal films are studied via AFM. In figure 6.8 topographies of the aluminum (left topography) and tantalum (right topography) films are shown for the 10 nm thick samples. From these AFM-topographies one can exclude any overlaid structure on a $2\,\mu m$ scale. However, one can see that the aluminum film shows a higher roughness than the tantalum film, as I saw earlier as well (see [82,84], and sections 4.3 and 4.4). This is interesting since the tantalum film's resistivity is much higher than the aluminum film's resistivity.

6.3.3.2 Electrochemical anodic oxidation of the films

Electrochemical anodic oxidation of polycrystalline aluminum films was already discussed in detail in section 4.3, [84] therefore the main focus is now set to the electrochemical oxidation of the amorphous tantalum films.

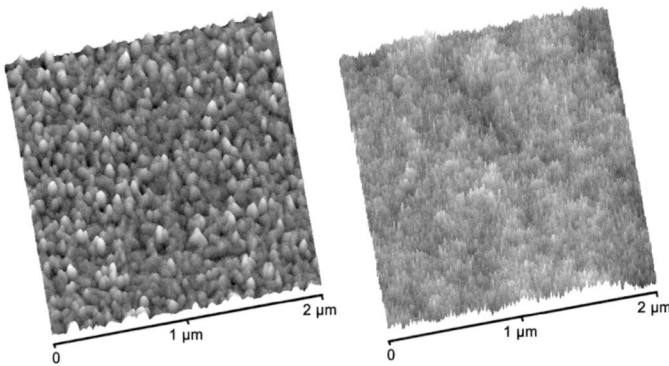

Figure 6.8: **Left:** Topography of the aluminum film on glass. **Right:** Topography of the tantalum film on glass.

The flowed charge during the tantalum oxidation cyclovoltammograms is integrated and plotted in figure 6.9 as the flowed charge, respectively the summed charge of the subsequently carried out oxidations versus its oxidation potential. After the first oxidations a nearly constant oxidation charge is measured, compare the linear slope of the summed charge. Therefore one can conclude that the electrochemical oxidation of the amorphous tantalum as a linear dependence on the oxidation potential. What means that an uniform oxidation occurs (crosschecked via XPS studies [212]).

When oxidizing with voltages $>$ 8.5 V (9 V and 9.5 V) one can clearly see a decrease in the flown charge. (For other samples with thinner metal films one can observe a decrease in the flown charge at lower voltages, which is not presented here). Meaning less tantalum metal is oxidized and consumed to form the oxide. But for defined voltage values one has to keep in mind that there is a big voltage drop for the last oxidations with the thicker oxide. Due to the high resistance of the remaining metal film, this would lead to a smaller amount of oxide formed. Therefore care was taken when the thickness of the remaining metal film was derived (see below).

The exchanged charge is due to the exchange current of the oxidation. From these changes a determination of the formed oxide and therefore of the remaining metal thickness should be possible. One could think of a direct calculation of the flown exchange current and its related amount of tantalum oxide being formed. But problems can occur due to voltage drops in the thin remaining metal films. However, in a first approach one would get a tantalum film oxide forma-

6.3 About the stability of thin Al–AlOx and Ta–TaOx interfaces

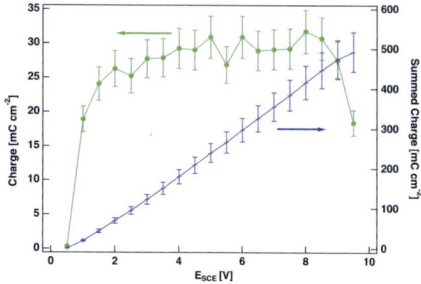

Figure 6.9: The flowed charge during the cyclovoltammograms is integrated and plotted here as the flowed charge, respectively the summed charge of the subsequent oxidations versus its oxidation potential.

tion factor of 1.8 nm/V. To shed more light on the oxide thickness we additionally performed XPS-sputter profile studies. We found an oxide film formation factor of 2.14 nm/V [212], what is in good agreement with literature values (1.3 - 2.4 nm/V) [121].

From this tantalum oxide film formation factor one derives easily via the ratio of the densities of tantalum (ρ_{Ta} = 16.6 g/cm^3 [237]) and tantalum–oxide ($\rho_{Ta_2O_5}$ = 8.73 g/cm^3 [237]) a tantalum film consumption factor of $\approx$ 1 nm/V. Comparably the aluminum film consumption factor was determined to $\approx$ 1.2 nm/V from the determined aluminum oxide film formation factor of 1.8 nm/V (literature values of 0.75 - 2 nm/V [121] for aluminum oxide film formation factor).

With these film consumption factors one can then determine the oxide's underlying metal film's thickness and study its electrical resistivity as it is done in the following section.

6.3.3.3 Electrical resistivity of thin aluminum and tantalum films

We already know that our aluminum films are polycrystalline (deduced from section 4.3, [84]) and our tantalum films are amorphous (compare section 4.4, [82]). Therefore one has to be aware of the different conduction mechanisms and different electron pathways in the films. The films were thinned via the electrochemical anodic oxidation, to avoid any preparation effects when very thin films are prepared just by evaporating different thick films. Then one can easily measure very thin film's resistivities via the oxide's underlying metal film's resistivity (see drawing in figure 6.7 and the results in figure 6.10).

The thickness dependent resistivities for the whole aluminum and tantalum

6.3.3 Discussion

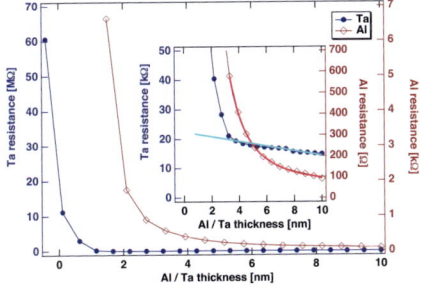

Figure 6.10: Electrical resistance of the tantalum and aluminum film (underlying the oxide) in dependence of the metal films thickness. The inset shows a zoom. The linear fit between 3.2 nm and 10 nm for the tantalum is marked as a blue line, whereas the power law fit between 3.3 nm and 10 nm for the aluminum resistance values is marked in red.

films with the $0.04\,\text{cm}^2$ oxide covered area show differences between aluminum and tantalum films (compare figure 6.10). For tantalum a linear increase of the resistance R with decreasing metal thickness d is found between 3.2 nm and 10 nm. A light blue marked linear fit is added in figure 6.10 for tantalum. This linear fit has a slope of $R = 22441\,\Omega - 795\,\Omega\cdot\text{nm}^{-1}\cdot d$ with an G value, for the goodness of the linear fit, of 0.96. This means that the resistance changes linear with the thinner film thickness up to a value of ≈ 3 nm.

Whereas the aluminum film show the power law like slope, as it is shown in [403]. The power law fit (with fit parameters of $R = 61\,\Omega + 9504\,\Omega\cdot\text{nm}^{-1}\cdot d^{-2.4}$) is added in figure 6.10 as a red curve.

In addition I want to emphasize that all resistance data for tantalum show nearly the same values after waiting long periods, what was crosschecked with other samples lying in the shelf for months. In general this stability of the tantalum–tantalum oxide interface is interesting. For aluminum this is quite different. Previously, regarding the aluminum–aluminum oxide interface of thin aluminum films I found that that up to one monolayer of the base aluminum electrode can be oxidized during 10^5 s after the end of potentiostatic oxidation just due to field strength dependent migration processes in anodic oxides, even in Ultra High Vacuum, (see section 4.3 [84]). Therefore only the data are plotted for a thickness of the remaining metal film up to 1.5 nm.

To gain more insight into these resistivity data we determined the correspond-

6.3 About the stability of thin Al–AlOx and Ta–TaOx interfaces

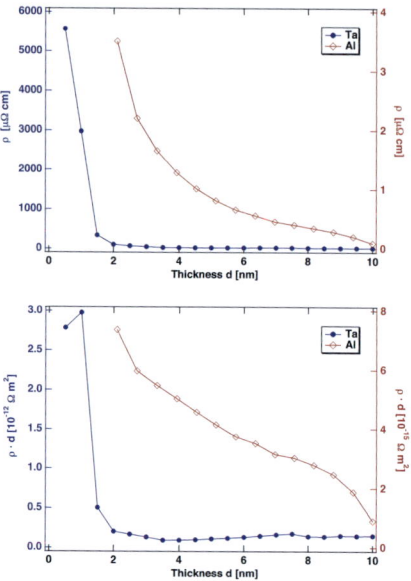

Figure 6.11: **Upper graph:** specific resistivity ρ as a function of the film thickness d for tantalum and aluminum. **Lower graph:** product of the specific resistivity ρ and film thickness d as a function of d for tantalum and aluminum. Here just the area of 0.04 cm² of the metal film under the oxide is taken as a basis.

ing specific resistivity ρ of the samples where only the 0.04 cm² of the metal film under the oxide is taken as a basis. This specific resistivity ρ is plotted in the upper graph of figure 6.11 as a function of the film thickness d for tantalum and aluminum. Additionally the product of the specific resistivity ρ and film thickness d is plotted in the lower graph of figure 6.11 as a function of d. Here the linear (Ta) and power law (Al) like slope can be seen as well. This is on the first glance quite surprising, since one would expect in general a linear dependence that could be accounted for with a simple Fuchs–Sondheimer approach [203, 224]. However, we already showed that this is not appropriate for the very thin metal films (see section 4.4, [82]). One can find some further literature presenting as well data that are not describable with a simple Fuchs–Sondheimer approach, regarding a thickness dependent conductivity study of thin metal films [220]. In [220] an 'U' shaped $\rho \cdot d$ versus d plot is presented. My data lie on the thin film thickness side of the presented 'U' shaped $\rho \cdot d$ versus d plot, what is one more evidence, that my experimental data are not describable with a simple Fuchs–Sondheimer approach [203, 224].

The experiments presented here are the first time that the resistivity is explored as a function of the metal layer thickness by a metal consumption experiment. One can clearly see that the traditional models of the electrical resistivity cannot be simply transferred to the metal–metal oxide systems.

Furthermore the difference between aluminum and tantalum is interesting. Thus one has to be aware of the electron pathways in the metal films. In our polycrystalline aluminum films we have a much longer elastic mean free path of the electrons $\approx$ 50 Å [404, 405] than for the amorphous tantalum film $\approx$ 5 Å (see section 4.5, [231]).

Therefore the thinning of the metal film has a stronger influence on the scattering of the electrons for the aluminum metal as for the tantalum. This is schematically added in figure 6.7 where possible electron pathways are marked. The scattering at the bottom side of the metal oxide has a much more pronounced effect for aluminum than for tantalum. For tantalum the big changes occur just below a metal thickness of $\approx$ 3 nm. This is the thickness where around 7 layers tantalum are present underneath the tantalum oxide [237]. Just then the additional scattering at the metal–oxide interface has an effect. This finding was crosschecked with a 5 nm thick tantalum film. Here at around $\approx$ 3 nm remaining metal film thickness significant changes can be observed as well (not presented here).

Based on these measurements and taking into account our studies on the amorphous tantalum films (see sections 4.4, [82] and 4.5, [231]), I explain the big difference between aluminum and tantalum resistivities and their stabilities with their different kind of interfaces. In the case of tantalum one has to consider an amorphous (metal) – amorphous (oxide) interface, whereas for aluminum a polycrystalline (metal) – amorphous (oxide) interface exists (section 4.5, [212, 231]). The latter one is not stable due to migration processes along the grain boundaries etc, enabling the field strength dependent migration processes after the end of the potentiostatic oxidation of the polycrystalline aluminum.

6.3.4 Conclusion

Thin aluminum and tantalum films were studied using AFM, electrochemical anodic oxidation and their thickness dependent resistivity. Thereby anodic oxidation was carried out to enable an easy preparation of very thin metal layer thicknesses. A rather linear response of the oxide thickness formation to the applied electrochemical oxidation potential was found. Resistivity studies regarding the

6.3 About the stability of thin Al–AlOx and Ta–TaOx interfaces

metal and oxide's stability using electrochemically prepared metal–metal oxide interfaces showed that tantalum exhibits a more stable interface than aluminum. This was attributed to the amorphous character of the films in case of tantalum. Whereas the aluminum–aluminum oxide with its polycrystalline–amorphous interface leads to less stable film systems.

The needs of applications where such a stable metal–metal oxide interface is necessary can be meet by tantalum–tantalum oxide systems. One example where this is already exploited is the good temperature stability of thin anodic oxide films in metal–metal oxide–metal structures [85]. Further examples that exploit the virtue of this interface's stability will show up in the future.

7 Appendix

7.1 Refereed Papers published

- Detlef Diesing, Domocos Kovacs, Kevin Stella and Christian Heuser, "Characterization of atom and ion-induced 'internal' electron emission by thin film tunnel junctions", Nuclear Instruments and Methods in Physics Research Section B: Beam Interactions with Materials and Atoms, in press (2011)

- Kevin Stella, Domocos A. Kovacs, Wolfgang Brezna, Jürgen Smoliner and Detlef Diesing, "Charge transport through thin amorphous titanium and tantalum oxide layers", J. Electrochem. Soc. **158**, P65 (2011)

- Kevin Stella, Damian Bürstel, Eckart Hasselbrink and Detlef Diesing, "Thin tantalum films on crystalline silicon - a metallic glass", phys. stat. solidi RRL **5**, 68 (2011)

- Kevin Stella, Domocos A. Kovacs and Detlef Diesing, "Photosensitive metal--insulator–semiconductor devices with stepped insulating layer", Electrochem. Solid-State Lett. **12**, H453 (2009)

- Kevin Stella, Damian Bürstel, Steffen Franzka, Detlef Diesing and Oliver Posth, "Preparation and properties of thin amorphous tantalum films formed by small e-beam evaporators", J. Phys. D. **42** 135417 (2009)

- Kevin Stella and Detlef Diesing, "The Field Dependence of Aging Processes: Ion Migration in Anodic Oxide Films after Potentiostatic Formation", J. Electrochem. Soc., **154**, C 663 - C 670 (2007)

7.2 Refereed Papers in preparation

- James E. Sadler, Doug S. Szumski, Agnieska Kierzkowska, Samantha R. Catarelli, Kevin Stella, Richard Nichols, Mariano Fonticelli, Guillermo Benitez, Bárbara Blum, Roberto C. Salvarezza, Walther Schwarzacher, "An In-Situ Electrochemical Method for the Formation of a Layer of Octanethiol on Electro-deposited Nickel", submitted to J. Phys. Chem.

- Kevin Stella, Domocos A. Kovacs, Wolfgang Brezna, Jürgen Smoliner and Detlef Diesing, "Transport of excited holes through metal–insulator–semiconductor devices", to be submitted to Phys. Rev. B

- Kevin Stella, Steffen Franzka and Detlef Diesing, "About the stability of thin Al–AlOx and Ta–TaOx interfaces ", to be submitted

- Katrin Bruder, Achim W. Hassel, Yanka Jeliazova, Kevin Stella, and Detlef Diesing, "Temperature dependence of chemoelectronic devices", to be submitted

- Kevin Stella, P. Thissen, Eckart Hasselbrink and Detlef Diesing, "Internal photoemission studies on the temperature dependence of TaTaOxAu chemoelectronic sensors ", in preparation

- Kevin Stella and Detlef Diesing, "Temperature dependence of stepped-MIS chemoelectronic sensors", in preparation

- Kevin Stella, Detlef Diesing and Eckart Hasselbrink, "Molecular reaction chemicurrent studies with MIS sensors", in preparation

7.3 Invited talk

- 250th ACS National Meeting, American Chemical Society
  Boston (USA), 22. – 26. Aug. 2010
  "Chemical experiments with pulsed atomic and molecular beams on the catalytic active surfaces of semiconductor heterosystems"

7.4 Poster and talks

- DPG Frühjahrstagung der Sektion Kondensierte Materie
  Dresden (Germany), 13. – 18. March 2011
  "Monitoring particle and photo induced electronic excitations by metal–insulator–semiconductor devices"

- DPG Frühjahrstagung der Sektion Kondensierte Materie
  Dresden (Germany), 13. – 18. March 2011
  "Nanometer thin tantalum oxide capacitors: Characterization of temperature stability and built in electric fields "

- Energy Dissipation at Surfaces – 4th International Workshop SFB 616
  Kloster Schöntal (Germany), 5. – 8. Sept. 2010
  "Stepped metal–insulator–semiconductor heterosystems for surface chemistry studies"

- 250th ACS National Meeting, American Chemical Society
  Boston (USA), 22. – 26. Aug. 2010
  "Surface chemistry with photo-sensitive stepped metal–insulator–semiconductor heterosystems"

- 217th ECS Meeting, The Electrochemical Society
  Vancouver (Canada), 25. – 30. April 2010
  "Electrochemically prepared photo-sensitive metal-insulator-semiconductor devices with stepped insulating layer"

- DPG Frühjahrstagung der Sektion Kondensierte Materie
  Regensburg (Germany), 21. – 26. March 2010
  "Preparation and optical properties of metal-insulator-metal devices based on Ti and Ta for photocurrent and chemicurrent applications"

7.4 Poster and talks

- DPG Frühjahrstagung der Sektion Kondensierte Materie
  Regensburg (Germany), 21. – 26. March 2010
  "Preparation and properties of thin amorphous tantalum films formed by small e-beam evaporators"

- Energie Dissipation an Oberflächen SFB 616 Summer School
  Essen (Germany), 26. Sept. – 1. Okt. 2009
  "Development of a sensor for the detection of electronic dissipation processes induced by chemical reactions at surfaces"

- Energie Dissipation an Oberflächen SFB 616 Workshop
  Remagen (Germany), 6. – 9. Sept. 2009
  "Virtue of electrochemically prepared metal-insulator-semiconductor devices with stepped insulator layers"

- SFB 616 Mitarbeiterworkshop 2009, SFB 616
  Papenburg (Germany), 17. - 19. June 2009
  "Physik und Chemie an Metall-Isolator-Metall- (MIM) und Metall-Isolator-Halbleiter- (MIS) Bauteilen"

- 108. Hauptversammlung der Deutschen Bunsen-Gesellschaft für Physikalische Chemie e.V., Bunsentagung 2009
  Köln (Germany), 21. - 23. May 2009
  "Electrochemically prepared metal-insulator-semiconductor devices with stepped insulator layers"

- 11th JCF-Frühjahrssymposium
  Essen (Germany), 11. - 14. March 2009
  "Metal-insulator-semiconductor devices with electrochemically prepared stepped insulator layers "

- Energy Dissipation at Surfaces – 3rd International Workshop SFB 616
  Bad Honnef (Germany), 25. - 28. Aug. 2008
  "Device types for chemicurrent detection: Metal-insulator-metal and stepped-metal-insulator-semiconductor"

- 3rd Euregio Workshop – AGEF Seminar 'Interfacial Electrochemistry'
  Kerkrade (The Netherlands), 2. – 3. June 2008
  "Electrochemical preparation of electronic devices"

- SFB 616 Mitarbeiterworkshop 2008, SFB 616
  Papenburg (Germany), 19. – 21. May 2008
  "Molecular beams and electronic devices in chemicurrent experiments"

- 72. Jahrestagung der Deutschen Physikalischen Gesellschaft
  Berlin (Germany), 25. – 29. Feb. 2008
  "The field dependence of ageing processes: Ion migration in amorphous aluminum and tantalum oxide films after potentiostatic formation"

Curriculum vitae

Name	Stella, Kevin
Geburtsdatum	15.04.1984
Geburtsort	Duisburg
Nationalität	deutsch

Doktorand an der Universität Duisburg-Essen
Juni 2008 – Juni 2011
Promotionsstipendiat der Studienstiftung des deutschen Volkes
Chemie Studium an der Universtät Duisburg-Essen, University of Liverpool
Oktober 2003 – April 2008

- Diplomarbeit: Prof. Dr. Eckart Hasselbrink, "Elektronische Anregungsprozesse bei der Adsorption von Molekülen auf Platin und Goldoberflächen", Oktober 2007 – April 2008

- Forschungsaufenthalt: Prof. Dr. Richard J. Nichols, Department of Chemistry, University of Liverpool, April - Juli 2007

- Studienstipendiat der Studienstiftung des deutschen Volkes
Februar 2006 – April 2008

- Sommersemester 2003 als "Vorsemester": Physik Vordiplom, Physik Praktikum, Mathematik für Chemiker 1 und 2, Gefahrstoffrechtskunde, Toxikologie, Informatik

Wehrersatzdienst	3. 6. 2002 - 31. 3. 2003
Reifeprüfung	Sophie-Scholl-Gymnasium
	Oberhausen-Sterkrade (28. Juni 2002)
Schulbildung	4 Jahre Grundschule, 8 Jahre Gymnasium
	(Jgst. 11 übersprungen)

Teilnehmer des vierteiligen Naturwissenschaftlichen Kollegs der Studienstiftung in der Arbeitsgruppe Quantenoptik

Jugend forscht, Fachbereich Physik, "Energieverluste im Stand–by–Betrieb" 1997

GDCh–Jungchemikerforum Duisburg–Essen
1. Sprecher 2008 – 2010, 2. Sprecher 2007 – 2008, Ausrichtung des 11. Internationalen JCF-Frühjahrssymposiums in Essen, März 2009

Erklärung

Hiermit versichere ich, dass ich die vorliegende Arbeit mit dem Titel

"Electronic dissipation processes during chemical reactions on surfaces"

selbst verfasst und keine außer den angegebenen Hilfsmitteln und Quellen benutzt habe, ich bisher in keinem Promotionsverfahren gescheitert bin und dass die Arbeit in dieser oder ähnlicher Form noch bei keiner anderen Universität eingereicht wurde.

Essen, im Juni 2011

Danksagung

Ich bedanke mich bei Herrn Prof. Dr. Eckart Hasselbrink für die interessante Themenstellung und die frühe Aufnahme in den Arbeitskreis.

Mein besonderer Dank gilt Herrn Dr. Detlef Diesing für die Vermittlung der Begeisterung für die Chemoelektronik, die Hilfe bei der Durchführung der Messungen, der fachlichen Unterstützung, Schreibstunden an buchartigen Papern, und den zahlreichen Diskussionen.

Bei Damian Bürstel, Domocos A. Kovacs, Ievgen Nedrygailov und Michael Scheele möchte ich mich für die sehr gute Zusammenarbeit bedanken. Ievgen Nedrygailov und Jan Weber danke ich außerdem für die sprachlichen Korrekturvorschläge. Dr. Steffen Franzka danke ich für die AFM Messungen. Weiterer Dank gebührt Elke Norman, Hans Vanheiden, Jürgen Leistikov und Dirk Gründer für die Hilfe bei technischen Problemen. Des Weiteren danke ich allen Mitgliedern des Arbeitskreises für die freundliche Arbeitsatmosphäre und stets gewährte Unterstützung.

Bei Dr. Wolfgang Brezna und Prof. Dr. Jürgen Smoliner möchte ich mich für die gute Wiener Zusammenarbeit bedanken.

Genauso danke ich dem Arbeitskreis Prof. Dr. Andreas Wucher für gute Zusammenarbeit und Bereitstellung von Räumlichkeiten in Duisburg.

Für die XPS-Sputter Messungen und Samstagsdiskussionen am Max-Planck-Institut für Eisenforschung gebührt mein Dank Katrin Bruder und Prof. Dr. Achim W. Hassel, Universität Linz.

Ebenso danke ich Prof. Dr. Günter Dumpich für stimulierende Diskussionen.

Für die SEM Messungen möchte ich mich bei Oliver Posth und Christian Wirtz aus dem Arbeitskreis Prof. Dr. Michael Farle bedanken.

Jürgen Gündel-Graber, Prof. Dr. Roland Boese und Prof. Dr. Matthias Epple aus der Anorganischen Chemie danke ich für die Röntgendiffraktometrie Messungen und Diskussionen.

Dem Institut für Bio- und Nanosysteme 2 des Forschungszentrums Jülich danke ich für die Hilfe bei der Herstellung dünner Titan- und Tantalfilme, sowie der Werkstatt für die kniffligen Anfertigungen.

Auch möchte ich der DFG danken die den SFB 616 Energiedissipation an Oberflächen so gut unterstützt, dass die zur Durchführung dieser Arbeit nötigen Apparaturen zur Verfügung standen.

Bei Prof. Dr. Marika Schleberger, Prof. Dr Eckart Hasselbrink und Prof. Dr. Achim W. Hassel möchte ich mich auch ausdrücklich für die Begutachtung dieser

Arbeit bedanken. Prof. Dr. Mathias Ulbricht danke ich für die Übernahme des Vorsitzes der Prüfungskommission.

Ebenso danke ich der Studienstiftung des deutschen Volkes für die gewährten Stipendien, die sowohl zur Durchführung des Studiums und der Promotion, als auch zur Bewusstseinserweiterung in vielerlei Hinsicht äußerst förderlich waren.

Besonders bedanken möchte ich mich auch bei meinen Eltern Karin und Klaus Stella, meiner Schwester Kelly und meinen Großeltern, die mich während des gesamten Studiums unterstützt haben. Durch die Vollendung dieser Arbeit hoffe ich einen Teil davon zurückgeben zu können.

Der disserta Verlag bietet die kostenlose Publikation
Ihrer Dissertation als hochwertige
Hardcover- oder Paperback-Ausgabe.

Fachautoren bietet der disserta Verlag
die kostenlose Veröffentlichung professioneller Fachbücher.

Der disserta Verlag ist Partner für die Veröffentlichung
von Schriftenreihen aus Hochschule und Wissenschaft.

Weitere Informationen auf www.disserta-verlag.de

disserta
―――
Verlag